WITHDRAWN BY THE
UNIVERSITY OF MICHIGAN

MONOGRAPHS ON
APPLIED PROBABILITY AND STATISTICS

General Editor
D.R. COX, F.R.S.

OPTIMAL DESIGN
AN INTRODUCTION TO THE
THEORY FOR PARAMETER ESTIMATION

Optimal Design

AN INTRODUCTION TO THE
THEORY FOR PARAMETER ESTIMATION

S.D. SILVEY

Professor of Statistics
University of Glasgow

1980
LONDON   NEW YORK
CHAPMAN AND HALL
150TH ANNIVERSARY

First published 1980 by
Chapman and Hall Ltd
11 New Fetter Lane, London EC4P 4EE

Published in the USA by
Chapman and Hall
in association with Methuen, Inc.
733 Third Avenue, New York, NY 10017

© 1980 S.D. Silvey

Printed in Great Britain at the
University Press, Cambridge

ISBN 0 412 22910 2

All rights reserved. No part of this book may be reprinted, or
reproduced or utilized in any form or by any electronic,
mechanical or other means, now known or hereafter invented,
including photocopying and recording, or in any information
storage and retrieval system, without permission in writing
from the publisher.

British Library Cataloguing in Publication Data

Silvey, Samuel David
  Optimal design—(Monographs on applied probability and
  statistics)
  1. Design—Mathematics   2. Regression analysis
  I. Title   II. Series
  519.5'36   QA278.2   80-40618

ISBN 0-412-22910-2

CONTENTS

	Preface	page vii
1	Introduction	1
1.1	*Historical note*	1
1.2	*The general problem*	2
1.3–1.6	*Examples*	4
1.7	*Discriminating among models*	8
2	Linear theory	9
2.1	*Definition*	9
2.2	*Design criteria*	10
2.3	*A property of criterion functions*	13
2.4	*Approximate theory*	13
3	Approximate theory for linear regression design	15
3.1	*The general problem*	15
3.2	*The set of information matrices*	15
3.3	*The criterion function*	16
3.4	*The solution set of design measures*	17
3.5	*Directional derivatives*	17
3.6, 3.7	*Basic theorems*	19
3.8	*Example: quadratic regression with a single control variable*	20
3.9	*Theorem: equivalence of D- and G-optimality*	22
3.10	*Corollary*	22
3.11–3.12	*Discussion*	23
3.13	*Optimal design measures with singular information matrices*	25
4	Algorithms	28
4.1	*Introduction*	28
4.2	*Design measure algorithms*	29
4.3	*The V-algorithm for D-optimality*	30

4.4	*An example of the use of the W-algorithm*	32
4.5	*General comments*	33
4.6	*Convergence considerations*	35
4.7	*Review*	36
4.8	*N-observation designs*	37
5	Approximate theory—particular criteria	40
5.1	*D-optimality*	40
5.2	*D_A- and D_S-optimality*	43
5.3	*A linear criterion function*	48
5.4	*c-optimality*	49
5.5	*Examples*	49
5.6	*Other criteria*	52
6	Non-linear problems	53
6.1	*Introduction*	53
6.2–6.6	*Examples*	54
7	Sequential designs	62
7.1	*Objective*	62
7.2	*An alternative method*	63
7.3	*Convergence considerations*	64
7.4	*Inference from sequentially constructed designs*	66

APPENDIX

A.1	*Concavity results*	69
A.2	*Carathéodory's Theorem*	72
A.3	*Differentiability*	74
A.4	*Lagrangian theory: duality*	78

PREFACE

Prior to the 1970's a substantial literature had accumulated on the theory of optimal design, particularly of optimal linear regression design. To a certain extent the study of the subject had been piecemeal, different criteria of optimality having been studied separately. Also to a certain extent the topic was regarded as being largely of theoretical interest and as having little value for the practising statistician. However during this decade two significant developments occurred. It was observed that the various different optimality criteria had several mathematical properties in common; and general algorithms for constructing optimal design measures were developed. From the first of these there emerged a general theory of remarkable simplicity and the second at least raised the possibility that the theory would have more practical value.

With respect to the second point there does remain a limiting factor as far as designs that are optimal for parameter estimation are concerned, and this is that the theory assumes that the model underlying data to be collected is known *a priori*. This of course is seldom the case in practice and it often happens that designs which are optimal for parameter estimation allow no possibility of model validation. For this reason the theory of design for parameter estimation may well have to be combined with a theory of model validation before its practical potential is fully realized.

Nevertheless discussion in this monograph is limited to the theory of design optimal for parameter estimation. I have not attempted to give a comprehensive account even of this theory, but merely to present in a general way what I consider to be its more important aspects. There are many omissions, but it is my hope that the monograph will facilitate study of the subject by providing a structure into which most of the particular results in the literature can be fitted.

There are two approaches to the approximate theory of linear regression design; one uses differential calculus, the other duality

PREFACE

theory. I have concentrated on the former though recent results suggest that duality theory may be the more powerful tool, since the calculus approach runs into difficulties when the criterion function is not differentiable at its optimum. I have given some indication, though, of how duality theory can be used.

Roughly two-thirds of the book is devoted to linear theory and the remainder to non-linear. The former is now firmly established and there are few outstanding problems connected with it. The same cannot be said of non-linear theory. Much empirical and theoretical work remains to be carried out before the behaviour of general methods of constructing designs for non-linear estimation problems is fully understood.

In order to make the monograph self-contained an appendix on relevant mathematical theory is included. Results on differentiability and a proof of Carathéodory's Theorem I have gleaned from Rockafeller's book entitled Convex Analysis.

Finally I wish to express my thanks to Miss Mary Nisbet who typed from a not very legible manuscript with her usual speed, accuracy and good humour.

Glasgow, February, 1980   S.D. SILVEY

1
INTRODUCTION

1.1 *Historical note*

The theory with which we shall be concerned in this monograph has its origins in a paper by Smith (1918). Little development occurred until the 1950's when the subject was taken up by various authors including Elfving (1952, 1955, 1959), Chernoff (1953), de la Garza (1954, 1956), Guest (1958) and Kiefer (1958, 1959), with whose name it is particularly associated. During the 1960's it was pursued vigorously by him and other authors including Hoel (1961, 1965), Karlin & Studden (1966) and Atwood (1969). In the early 1970's the core of the theory was crystallized in papers by Fedorov & Malyutov (1972), Whittle (1973) and Kiefer (1974).

Meantime there were two parallel developments. One of these is associated particularly with the name of G.E.P. Box who was concerned more with developing methods for tackling applied problems than with general mathematical theory. This work is reported in papers by Box & Wilson (1951), Box & Lucas (1959), Box & Draper (1959), Box & Hunter (1965a, 1965b), Hill & Hunter (1974) and Hunter, Hill & Henson (1969). Not surprisingly, although their aims were different there is considerable overlap in ideas between what might be called the Kiefer and the Box approaches, and this overlap has become more apparent in recent years when there has been more emphasis on developing tools for applying the Kiefer-type theory, a seminal paper in this context being that of Wynn (1970).

The other development took place in the USSR and is associated particularly with the name of Fedorov. In it both the mathematical theory and algorithms for applying it were studied. The earlier part of the work is reported in the English literature in a book by Fedorov (1972). Since the publication of this book there have been further advances in the theory and it is the object of this monograph to provide a concise account of what, in the author's view, are the more important aspects of the theory as it exists in the late 1970's.

The list of authors mentioned above is by no means comprehensive. For a recent bibliography, see Ash & Hedayat (1978).

1.2 The general problem

We start by considering a rather general problem.

The distribution of an observable random variable $\tilde{y}$ depends on:

(i) a column vector $u = (u_1 \ u_2 \ \ldots \ u_r)^T$ of real variables called control variables because they can be chosen by the experimenter;

(ii) a column vector $\theta = (\theta_1 \ \theta_2 \ \ldots \ \theta_k)^T$ of parameters which are fixed but unknown to the experimenter; these or certain functions of them are of interest to him;

(iii) a column vector $\tau = (\tau_1 \ \tau_2 \ \ldots \ \tau_l)^T$ of nuisance parameters; these also are fixed and unknown but they are not of primary interest.

The distinction between the parameter θ of interest and the nuisance parameter τ is somewhat fuzzy, particularly when interest is centred on fewer than k functions of θ. Typically, though, the vector θ is involved, along with u, in $E(\tilde{y})$ and the vector τ with 'error variance'.

Vectors u can be chosen from a given set $\mathcal{U}$ in R^r; the true θ is known to belong to a set Θ in R^k and the true τ to a set Υ in R^l.

We shall assume that, for given u, θ, τ the distribution of $\tilde{y}$ is determined by a probability density function $p(y|u, \theta, \tau)$ with respect to a fixed σ-finite measure. Typically either y is discrete and the measure is counting measure or y is continuous and the measure Lebesgue.

The experimenter is allowed to take N independent observations on $\tilde{y}$ at vectors $u_{(1)}, u_{(2)}, \ldots, u_{(N)}$ chosen by him from the set $\mathcal{U}$. We shall refer to such a choice of N vectors, not necessarily all distinct as an *N-observation design*. The basic problem is: what N-observation design should be chosen?

We might adopt a full decision-theoretic approach to this problem, as does Brooks (1972) in a particular case; see also Bandemer (1979); that is, introduce a utility function and a prior distribution for θ, τ; calculate the Bayes' worth of each N observation design; and choose one with a maximum Bayes' worth. However this approach, while straightforward in principle, is fraught with computational difficulties in practice and the theory with which we shall be concerned adopts a less formal approach, based on Fisher's information matrix. The rationale is as follows.

Suppose that we wish to estimate θ. If we assume sufficient regularity of the family $\{p(y|u, \theta, \tau); u \in \mathcal{U}, \theta \in \Theta, \tau \in \Upsilon\}$ then for a single observation on $\tilde{y}$ at the vector u, the partitioned Fisher information matrix for (θ, τ) is

$$J(u; \theta, \tau) = \begin{pmatrix} J_{\theta\theta}(u) & J_{\theta\tau}(u) \\ J_{\theta\tau}^T(u) & J_{\tau\tau}(u) \end{pmatrix}$$

where $J_{\theta\theta}(u)$ is the $k \times k$ matrix whose (i,j)th element is $E\{-\partial^2 \log p(\tilde{y}|u, \theta, \tau)/\partial \theta_i \partial \theta_j\}$ and $J_{\theta\tau}(u)$, $J_{\tau\tau}(u)$ are similarly defined. Also the information matrix for N independent observations taken respectively at $u_{(1)}, \ldots, u_{(N)}$ is

$$L(\mathbf{u}; \theta, \tau) = \sum_{i=1}^{N} J(u_{(i)}; \theta, \tau) = \begin{pmatrix} L_{\theta\theta}(\mathbf{u}) & L_{\theta\tau}(\mathbf{u}) \\ L_{\theta\tau}^T(\mathbf{u}) & L_{\tau\tau}(\mathbf{u}) \end{pmatrix},$$

where $\mathbf{u} = (u_{(1)}, \ldots, u_{(N)})$ and $L_{\theta\theta}(\mathbf{u}) = \sum_i J_{\theta\theta}(u_{(i)})$, with similar definitions for $L_{\theta\tau}(\mathbf{u})$ and $L_{\tau\tau}(\mathbf{u})$.

Assume that $\mathbf{u}$ is such that $L(\mathbf{u}; \theta, \tau)$ is non-singular. (We shall see later that we sometimes have to modify this assumption, but we ignore this complication in the meantime.) Then $L^{-1}(\mathbf{u}; \theta, \tau)$ provides a 'lower bound' for the variance matrix of an unbiased estimator of (θ, τ) and its leading $k \times k$ sub-matrix

$$V(\mathbf{u}; \theta, \tau) = \{L_{\theta\theta}(\mathbf{u}) - L_{\theta\tau}(\mathbf{u})L_{\tau\tau}^{-1}(\mathbf{u})L_{\theta\tau}^T(\mathbf{u})\}^{-1} \quad (1.2.1)$$

is a lower bound for the variance matrix of an unbiased estimator of θ. Note the dependence of this lower bound on the design $\mathbf{u}$. Note also that if N is large and we have sufficient regularity, the variance matrix of the maximum-likelihood estimator of θ is approximately $V(\mathbf{u}; \theta, \tau)$.

Now suppose that we are primarily interested in estimating real-valued functions $g_1(\theta), \ldots, g_s(\theta)$ instead of θ itself. Denote by g the vector-valued function $(g_1 \ \ldots \ g_s)^T$ and by $Dg(\theta)$ the $k \times s$ matrix with (i,j)th component $\partial g_j(\theta)/\partial \theta_i$. Again assuming sufficient regularity, given the design $\mathbf{u}$ a lower bound for the variance matrix of an unbiased estimator of $g(\theta)$ is

$$V_g(\mathbf{u};\theta,\tau) = \{Dg(\theta)\}^T V(\mathbf{u};\theta,\tau)\{Dg(\theta)\} \qquad (1.2.2)$$

and again for large N the variance matrix of the maximum-likelihood estimator of g is close to this lower bound. Note that (1.2.1) is a special case of (1.2.2) with $g_i(\theta) = \theta_i$, $i = 1, \ldots, k$, and $Dg(\theta) = I_k$, the $k \times k$ identity matrix. Hence we shall concentrate the immediately following discussion on (1.2.2).

The basic idea underlying much of our theory is that we should choose $\mathbf{u}$ to make (1.2.2) 'as small as possible', or alternatively to make its inverse 'as large as possible'. By doing so we are choosing $\mathbf{u}$ to get as much information as possible about $g(\theta)$, in a Fisherian sense. This information is 'retrievable' at least when N is large and we estimate by maximum-likelihood.

Of course this informal approach to what is basically a decision problem is open to criticism. Nevertheless it has a lot in common with much non-Bayesian statistical practice and it does allow some progress to be made.

The problem still seems formidable and two immediate questions arise:

(i) In precisely what sense do we wish to make the matrix $V_g(\mathbf{u};\theta,\tau)$ small? As we shall see there are various possibilities, depending on the context.

(ii) Can we find a $\mathbf{u}$, say $\mathbf{u}_*$, independent of θ and τ, which makes $V_g(\mathbf{u};\theta,\tau)$ small in the desired sense for all θ and τ? There is a class of problems, namely linear regression problems, for which $V_g(\mathbf{u};\theta,\tau)$ simplifies to such an extent that the answer to this question is 'yes'. For other problems where no such $\mathbf{u}_*$ exists we have to think again, but we postpone discussion of this until we have investigated the simpler situation since this investigation produces some ideas about the more complicated one.

The following examples indicate some of the possibilities.

1.3 *Example. Linear regression with a single control variable*

Here u is real, $\theta = (\theta_0\text{–}\theta_1)^T$ and τ is a positive number. For given u, θ and τ, $\tilde{y}$ is $N(\theta_0 + \theta_1 u, \tau)$; that is, we are concerned with a simple normal linear regression model.

Let us suppose, for illustrative purposes that $\mathscr{U}$ is $[-1, 1]$ and consider the design $\mathbf{u} = (u_{(1)}, \ldots, u_{(N)})$, where $-1 \leq u_{(i)} \leq 1$. It is a standard result that

$$V^{-1}(\mathbf{u}; \theta, \tau) = \frac{1}{\tau}\begin{pmatrix} N & \sum u_{(i)} \\ \sum u_{(i)} & \sum u_{(i)}^2 \end{pmatrix}.$$

This is an example where $V(\mathbf{u}; \theta, \tau)$ simplifies as indicated above and there is the possibility of finding a $\mathbf{u}_*$ which is 'best' for all θ and τ. The problem becomes that of choosing $\mathbf{u}$ to make the matrix

$$M(\mathbf{u}) = \begin{pmatrix} N & \sum u_{(i)} \\ \sum u_{(i)} & \sum u_{(i)}^2 \end{pmatrix}$$

'as large as possible', when we are interested in estimating θ itself.

Two comments seem worth making at this point:

(i) This simple example can be used to destroy a vain hope. The hope is that where we have a problem in which $V^{-1}(\mathbf{u}; \theta, \tau)$ simplifies in this kind of way, we may be able to find a $\mathbf{u}_*$ which is 'best' in the very strong sense that $M(\mathbf{u}_*) - M(\mathbf{u})$ is nonnegative definite for all possible $\mathbf{u}$. It is not difficult to verify that no such $\mathbf{u}_*$ exists for this example. This forces us to think in general about making some function of V^{-1} large, and we return to this point in Chapter 2.

(ii) In this case, even if we drop the assumption of normality of $\tilde{y}$, but retain the assumptions that $E(\tilde{y}|u, \theta, \tau) = \theta_0 + \theta_1 u$ and $\text{var}(\tilde{y}|u, \theta, \tau) = \tau$, then $V(\mathbf{u}; \theta, \tau)$ is still the variance matrix of the least-squares estimator of θ, and of course this estimator is the best linear unbiased estimator. Hence for problems of this nature, that is, linear regression problems with homoscedasticity, we may drop the normal assumption and proceed on a least-squares basis, retaining the same spirit as in the Fisher information matrix approach.

1.4 Example. A classical design problem

What is sometimes called classical design theory is concerned with the allocation of treatments to experimental units. It is not immediately obvious how this fits into our general framework, but the following simple example illustrates how it does.

Suppose that we are interested in the separate and combined

effects of two treatments T_1 and T_2 on some numerical characteristic y of experimental units. We assume, as is usual, that any treatment combination affects only the mean value of $\tilde{y}$. Denote by θ_1 the effect of T_1 alone; by θ_2 that of T_2 alone; and by $\theta_1 + \theta_2 + \theta_3$ that of T_1 and T_2 in combination. To any experimental unit we can do one of four things:

(i) apply nothing;
(ii) apply T_1 alone;
(iii) apply T_2 alone;
(iv) apply both T_1 and T_2.

We may code this by writing $(0, 0)$ for (i), $(1, 0)$ for (ii), $(0, 1)$ for (iii), and $(1, 1)$ for (iv), and interpret the experimenter's choice of action for an experimental unit as that of choosing values of two control variables u_1 and u_2 each of which can take only the values 0 and 1. So our design space $\mathcal{U}$ consists of four points. Finally we may write

$$E(\tilde{y}|u, \theta, \tau) = \theta_0 + \theta_1 u_1 + \theta_2 u_2 + \theta_3 u_1 u_2$$
$$\text{var}(\tilde{y}|u, \theta, \tau) = \tau,$$

and we have a linear regression problem of exactly the same character from a design point of view as that in Section 1.3.

1.5 *Example. Quadratic regression with a single control variable; interest in a non-linear parametric function*

Here u is real, $\theta = (\theta_0 \; \theta_1 \; \theta_2)^T$, θ_2 is known to be negative while τ is real and positive. For given u, θ, τ, $\tilde{y}$ is $N(\theta_0 + \theta_1 u + \theta_2 u^2, \tau)$. Suppose that we are interested in estimating the value of u where $E(\tilde{y})$ is maximal; that is, we wish to estimate $g(\theta) = -\theta_1/(2\theta_2)$.

In this case, for an N observation design $\mathbf{u}$,

$$V^{-1}(\mathbf{u}; \theta, \tau) = \frac{1}{\tau} \begin{pmatrix} N & \sum u_{(i)} & \sum u_{(i)}^2 \\ \sum u_{(i)} & \sum u_{(i)}^2 & \sum u_{(i)}^3 \\ \sum u_{(i)}^2 & \sum u_{(i)}^3 & \sum u_{(i)}^4 \end{pmatrix}$$

We have the same kind of simplification of V as in Section 1.3 because the model is linear in θ. Note, however, the difference caused by interest in the *non-linear* function $g(\theta)$. As in (1.2.2) we

INTRODUCTION

now wish to choose u to minimize

$$\{Dg(\theta)\}^T V(\mathbf{u}; \theta, \tau)\{Dg(\theta)\}, \qquad (1.5.1)$$

where $\{Dg(\theta)\}^T = \frac{1}{2}\theta_2^{-2}(-\theta_2 \quad \theta_1)$. Because of the dependence of (1.5.1) on θ, there may not exist a $\mathbf{u}_*$ which minimizes (1.5.1) for all θ and we have a problem of a different character from these described in Sections 1.3 and 1.4.

1.6 Example. Probit analysis: a non-normal, non-linear problem

A stimulus may be applied to subjects at various levels u which can be chosen from an interval $\mathscr{U}$. A subject may or may not respond to level u and the probability of response at this level can be expressed to a good approximation as $\Phi\{(u - \theta_1)/\theta_2\}$, where Φ is the $N(0,1)$ distribution function. Suppose that N subjects are available for experimentation. To what levels of the stimulus should they be subjected in order to estimate θ_1 and θ_2 as well as possible?

Define a random variable $\tilde{y}$ which takes the value 1 for response and 0 for non-response. Then

$$p(y|u, \theta) = [\Phi\{(u - \theta_1)/\theta_2\}]^y [1 - \Phi\{(u - \theta_1)/\theta_2\}]^{1-y}.$$

By a standard result, for a design $\mathbf{u} = (u_{(1)}, \ldots, u_{(N)})$ Fisher's information matrix for θ_1, θ_2 is

$$\begin{pmatrix} \sum_i \frac{1}{\Phi_i(1 - \Phi_i)} \left(\frac{\partial \Phi_i}{\partial \theta_1}\right)^2 & \sum_i \frac{1}{\Phi_i(1 - \Phi_i)} \frac{\partial \Phi_i}{\partial \theta_1} \frac{\partial \Phi_i}{\partial \theta_2} \\ \sum_i \frac{1}{\Phi_i(1 - \Phi_i)} \frac{\partial \Phi_i}{\partial \theta_1} \frac{\partial \Phi_i}{\partial \theta_2} & \sum_i \frac{1}{\Phi_i(1 - \Phi_i)} \left(\frac{\partial \Phi_i}{\partial \theta_2}\right)^2 \end{pmatrix},$$

where we have written Φ_i in place of $\Phi\{(u_{(i)} - \theta_1)/\theta_2\}$ for typographical brevity. In accordance with the principle we have adopted we wish to choose $\mathbf{u}$ to make this matrix large, in some sense. It clearly depends on θ and again we have the complication of the non-existence of a $\mathbf{u}_*$ which makes it large for all θ.

We introduce this example to give some intuitive content to the problem facing us in the non-linear case. Clearly we do not wish to experiment at values of u for which $\Phi\{(u - \theta_1)/\theta_2\}$ is close either to 0 or 1, because observations at such values of u will be uninformative about θ_1 and θ_2. However the values of u at which such proximity occurs depend on θ_1 and θ_2. When we do not know the true values of these parameters how can we know whether we are experimenting at uninformative values of u? It is clear that the only possible solution to the problem is some form of sequential experimentation where we choose levels for future subjects on the basis of the responses of previous subjects. Generally such sequential experimentation will be necessary in the non-linear case, if we have no prior knowledge of the unknown parameters.

1.7 *Discriminating among models*

Everything that we have considered up to this point has been based on the assumption that we know, apart from the values of a number of parameters, the model underlying any observations that we may make. This of course is seldom the case in practice. Rarely is the statistician confident that a particular parametric model underlies data. Sometimes he may be confident that one of several models will be adequate, but he is not sure which. For this reason considerable effort has been devoted in recent years to the problem of designing to discriminate among models, each of which contains the same control variables.

While this is clearly an important type of problem we shall not discuss it further in this monograph. For a recent comprehensive bibliography, see Pereira (1977).

2
LINEAR THEORY

2.1 Definition

By linear theory we mean theory concerned with problems where we get the same kind of simplification of $V_g(\mathbf{u}; \theta, \tau)$—see (1.2.2)—as in Examples 1.3 and 1.4. This simplification occurs when

(i) $E(\tilde{y}|u, \theta, \tau) = \theta_1 f_1(u) + \theta_2 f_2(u) + \ldots + \theta_k f_k(u)$,

$f_1, \ldots, f_k$ being known functions;

(ii) $\text{var}(\tilde{y}|u, \theta, \tau) = \tau$;

(iii) we are interested in estimating *linear* functions of θ.

In fact we can have slightly greater generality than this because we might replace (ii) by

(ii)′ $\text{var}(\tilde{y}|u, \theta, \tau) = \tau v(u)$,

where v is a known function of u. However this case can immediately be reduced to the simpler one by considering, instead of y, $y' = y/\{v(u)\}^{1/2}$ and instead of $f_i(u)$, $f'_i(u) = f_i(u)/\{v(u)\}^{1/2}$. So there is no essential loss of generality in restricting attention to the simpler case defined by (i), (ii) and (iii).

We have already noted that certainly in this kind of situation we can drop the assumption of normality of $\tilde{y}$ and work in terms of least-squares rather than maximum likelihood. This we shall do for linear theory. Note further that, subject to (i) and (ii), choosing a vector u in a design space $\mathcal{U}$ is equivalent to choosing a k-vector x in the induced design space $\mathcal{X} = f(\mathcal{U})$, where f is the vector-valued function $(f_1 \ \ldots \ f_k)^T$. Thus there is no loss of generality and considerable notational convenience in replacing (i) by

(i)′ $E(y|x, \theta, \tau) = x^T \theta$,

and thinking of an N-observation design as a choice of N-vectors $x_{(1)}, \ldots, x_{(N)}$ from the induced design space $\mathcal{X}$, a subset of R^k.

Now consider the N-observation design $\mathbf{x} = (x_{(1)}, \ldots, x_{(N)})$. If

OPTIMAL DESIGN

we assume that observations on $\tilde{y}$ are independent, or more generally, uncorrelated, then the variance matrix of the least-squares estimator $\hat{\theta}$ of θ arising from this design is

$$V(\mathbf{x};\tau) = \tau(\sum x_{(i)}x_{(i)}^T)^{-1},$$

when $\mathbf{x}$ admits estimation of θ, that is, when $\sum x_{(i)}x_{(i)}^T$ is non-singular. Because $V(\mathbf{x};\tau)$ does not depend on θ and because of the simple way it depends on τ, our problem becomes: choose $\mathbf{x}$ with $x_{(i)} \in \mathscr{X}$, $i = 1, \ldots, N$ to make

$$M(\mathbf{x}) = \sum x_{(i)}x_{(i)}^T$$

large, in some sense. We now consider various ways in which we might wish to make $M(\mathbf{x})$ large.

2.2 Design criteria

We have already seen, in Section 1.3, that there is little point in seeking a design $\mathbf{x}_*$ which is optimal in the very strong sense that $M(\mathbf{x}_*) - M(\mathbf{x})$ is non-negative definite for all $\mathbf{x}$. That example makes it clear that only in exceptional circumstances is such an $\mathbf{x}_*$ likely to exist. Accordingly we have to content ourselves with attempting to find an $\mathbf{x}_*$ which maximizes some real-valued function $\phi\{M(\mathbf{x})\}$. Various functions ϕ are suggested by practical considerations.

2.2.1 D-optimality.
If we assume normality of errors then in the linear context under discussion a confidence ellipsoid for θ, of given confidence coefficient and for a given residual sum of squares, arising from the design $\mathbf{x}$, has the form

$$\{\theta : (\theta - \hat{\theta})^T M(\mathbf{x})(\theta - \hat{\theta}) \leqslant \text{constant}\},$$

where $\hat{\theta}$ is the least-squares estimate of θ. The content of this ellipsoid is proportional to $\{\det M(\mathbf{x})\}^{-1/2}$. A natural design criterion is that of making this ellipsoid as small as possible, in other words that of maximizing $\det M(\mathbf{x})$.

This is the celebrated criterion of D-optimality, the most intensively studied of all design criteria.

2.2.2 D_A- and D_s-optimality.
Suppose that our primary interest is in certain linear combinations of $\theta_1, \ldots, \theta_k$, those s linear

combinations which are elements of the vector $A^T\theta$, where A^T is an $s \times k$ matrix of rank $s < k$. If $\mathbf{x}$ is a design for which $M(\mathbf{x})$ is non-singular, the variance matrix of the least-squares estimator of $A^T\theta$ is proportional to $A^T\{M(x)\}^{-1}A$ and by the same argument as in Section 2.2.1, a natural criterion is that of maximizing $\det[A^T\{M(x)\}^{-1}A]^{-1}$.

This is the criterion of D_A-optimality, so called by Sibson (1974).

The canonical version of this occurs when $A^T = (I_s \; 0)$, that is, when we are interested in the first s parameters $\theta_1, \ldots, \theta_s$ with $s < k$. In this case, if $M(\mathbf{x})$ is partitioned in the obvious way:

$$M(\mathbf{x}) = \begin{pmatrix} M_{11}(\mathbf{x}) & M_{12}(\mathbf{x}) \\ M_{12}^T(\mathbf{x}) & M_{22}(\mathbf{x}) \end{pmatrix}$$

then it is a matter of elementary algebra to show that $(A^T M(\mathbf{x})A)^{-1}$ can be expressed in the form

$$M_{11}(\mathbf{x}) - M_{12}(\mathbf{x})\{M_{22}(\mathbf{x})\}^{-1}M_{12}^T(\mathbf{x})$$

and our design criterion becomes that of choosing $\mathbf{x}$ to maximize the determinant of this matrix.

This is the criterion of D_s-optimality; see Karlin & Studden (1966), Atwood (1969), Silvey & Titterington (1973), among many references.

There is a hidden difficulty here which gives rise to considerable theoretical complications, as we shall see. Suppose that we are interested in $A^T\theta$. Then it is possible that a design which admits estimation of $A^T\theta$ but does not admit estimation of θ itself is better for estimating $A^T\theta$ than any other design. A somewhat artificial example makes this clear. Suppose that we have two control variables x_1 and x_2, that $E(\tilde{y}|x_1, x_2, \theta) = \theta_1 x_1 + \theta_2 x_2$ and that our interest is centred on θ_1. If the design space $\mathscr{X}$ is the triangle with vertices $(0, 0)$, $(1, 0)$ and $(0, 1)$ then it is intuitively clear that the N-observation design according to which we take all N observations at $(1, 0)$ must at least be a serious competitor for providing a best estimator of θ_1. For this design $\mathbf{x}$ we have

$$M(\mathbf{x}) = \begin{pmatrix} N & 0 \\ 0 & 0 \end{pmatrix},$$

which of course is singular.

OPTIMAL DESIGN

Because of this, in developing the theory of D_A-optimality we must take account of designs **x** such that $A^T\theta$ is estimable but $M(\mathbf{x})$ is singular. We return to this point and the difficulties to which it gives rise later.

2.2.3 *G-optimality*.

Our primary interest may lie in predicting $E(\tilde{y})$ over some region $\mathscr{C}$ of R^k. Suppose that a design **x** has nonsingular $M(\mathbf{x})$. For fixed $c \in \mathscr{C}$, the variance of the least-squares estimator of $c^T\theta$, associated with **x** is proportional to $c^T\{M(\mathbf{x})\}^{-1}c$. One possibility is that we may wish to choose **x** to minimize $\max_{c \in \mathscr{C}} c^T\{M(\mathbf{x})\}^{-1}c$.

When $\mathscr{C} = \mathscr{X}$, the design space itself, this minimax criterion has been called G-optimality; see for instance, Kiefer & Wolfowitz (1960).

2.2.4 *E-optimality*.

This is a particular case of the general one discussed in Section 2.2.3 when $\mathscr{C}$ is the unit sphere:

$$\mathscr{C} = \{c \in R^k : c^Tc = 1\}.$$

Choosing **x** to minimize $\max_{\{c:c^Tc=1\}} c^T\{M(\mathbf{x})\}^{-1}c$ is equivalent to choosing **x** to maximize the minimum eigenvalue of $M(\mathbf{x})$, a criterion termed E-optimality: see Kiefer (1974).

2.2.5 *A linear criterion function*.

Again thinking in terms of predicting $E(\tilde{y})$ over a region $\mathscr{C}$, we may wish to minimize some average over $\mathscr{C}$ of $c^T\{M(\mathbf{x})\}^{-1}c$. Suppose that we average with respect to a probability distribution μ on $\mathscr{C}$. Then we seek to minimize $\int_{\mathscr{C}} c^T\{M(\mathbf{x})\}^{-1}c\mu(dc)$. It is not difficult to show that this integral can be expressed as $\text{tr}[\{M(\mathbf{x})\}^{-1}B]$, where tr denotes trace and $B = \int_{\mathscr{C}} cc^T\mu(dc)$, a non-negative definite matrix.

This is a linear criterion function: see Fedorov (1972).

Note that if B has rank s it can be expressed in the form AA^T where A is a $k \times s$ matrix of rank s. Then this criterion function can be expressed in the form

$$\text{tr}[A^T\{M(\mathbf{x})\}^{-1}A],$$

which exhibits its relationship with D_A-optimality. Again if $s < k$, we face the same difficulty as with D_A-optimality; we have to consider the possibility that designs with singular M may be optimal.

LINEAR THEORY

2.2.6 *c-optimality.* This is a particular case of Section 2.2.5 when μ puts probability 1 at a particular c. Now our criterion becomes that of minimizing $c^T\{M(\mathbf{x})\}^{-1}c$, for a fixed c, and this criterion has been termed c-optimality; see, for instance, Elfving (1952).

2.3 A property of criterion functions

The above list of possible criteria is not comprehensive though it does include the main ones considered in the literature. Note however that only criteria arising from interest in *linear* functions of θ have been included. As indicated earlier, if we are interested in *non-linear* functions of θ, we have a problem of a different character.

Note also that we have been led naturally sometimes to a minimization, sometimes to a maximization, problem; but of course any minimization problem can be rephrased as a maximization one. To simplify discussion we shall invariably think in terms of the maximization problem so that, for example, in Section 2.2.6 we would take as criterion function $-c^T\{M(\mathbf{x})\}^{-1}c$, maximizing which is equivalent to minimizing $c^T\{M(x)\}^{-1}c$.

Finally in this section we note that after each problem has been rephrased as a maximization one, all the criterion functions ϕ that we have introduced possess a natural and desirable property: if $\mathbf{x}_1$ and $\mathbf{x}_2$ are two N-observation designs such that $\mathbf{x}_1$ is better than $\mathbf{x}_2$ in the strong sense that $M(\mathbf{x}_1) - M(\mathbf{x}_2)$ is non-negative definite then $\phi\{M(\mathbf{x}_1)\} \geqslant \phi\{M(\mathbf{x}_2)\}$. This is quite easily verified in all cases.

2.4 Approximate theory

When statistical considerations have been used to fix a criterion function $\phi\{M(\mathbf{x})\}$ we have reduced the problem in the linear case to one in numerical analysis. However because of the discrete nature of N observation designs and the consequent unpleasantness of the corresponding set $\mathscr{M}(N)$, say, of matrices $M(\mathbf{x})$ on which ϕ is defined, this numerical analysis problem is not amenable to solution by standard optimization techniques, particularly when N is not small. The situation is analogous to the much simpler one where we wish to maximize a function defined on the integers. Because of the discrete domain, calculus techniques cannot be exploited in the solution. A commonly used device for this simpler problem is to extend the definition of the function to all real numbers t, use calculus to find the number t_* where the maximum of the extended function occurs and argue that the maximum of the function over the integers will occur at an integer adjacent to t_*. This idea has been adapted for the design problem and leads to what Kiefer has termed the *approximate theory* of linear regression design.

We start by noting that all the criterion functions ϕ introduced above are such that, for a real fixed positive constant a,

$$\phi\{aM(\mathbf{x})\} = \text{a positive constant} \times \phi\{M(\mathbf{x})\}.$$

Hence an $\mathbf{x}_*$ which maximizes $\phi\{aM(\mathbf{x})\}$ also maximizes $\phi\{M(\mathbf{x})\}$. Now a typical N-observation design $\mathbf{x}$ will contain a number, n say, of *distinct* vectors $x_{(1)}, \ldots, x_{(n)}$ replicated respectively $r_1, \ldots, r_n$ times, where $\sum r_i = N$. We may then associate with this design a discrete probability distribution on $\mathscr{X}$, the distribution η_N which puts probability $p_i = r_i/N$ at $x_{(i)}, i = 1, \ldots, n$. Now let $\tilde{x}$ be a random vector with distribution η_N and define

$$M(\eta_N) = E(\tilde{x}\tilde{x}^T) = \sum p_i x_{(i)} x_{(i)}^T = N^{-1} M(\mathbf{x}),$$

and re-interpret the N-observation design problem as that of finding η_{N*}, a probability distribution corresponding to an N-observation design, to maximize $\phi\{M(\eta_N)\}$. This interpretation in itself does not ameliorate the problem. However we can extend the definition of M to the set H of *all* probability distributions on $\mathscr{X}$: if $\eta \in$ H and $\tilde{x}$ is a random vector with distribution η, simply define

$$M(\eta) = E(\tilde{x}\tilde{x}^T);$$

and similarly we may extend the definition of ϕ to the wider class $\mathscr{M} = \{M(\eta): \eta \in \mathrm{H}\}$ of matrices.

Now consider the problem of finding η_* to maximize $\phi\{M(\eta)\}$ over H. This is much more tractable since $\mathscr{M}$ is a set with much nicer properties than $\mathscr{M}_N$ and we have the opportunity of exploiting calculus techniques in solving it. If we can find an η_* to solve this problem then hopefully an N-observation design $\mathbf{x}_*$ whose associated probability distribution approximates η_* will be close to optimal for the N-observation design problem.

This neat idea leads to some elegant and useful theory which we shall discuss in the next chapter.

3

APPROXIMATE THEORY FOR LINEAR REGRESSION DESIGN

3.1 *The general problem*

The previous chapter has motivated consideration of the following problem.

Let $\mathscr{X}$ be a given compact subset of Euclidean k-space, to be thought of as an induced design space. To postulate that $\mathscr{X}$ is compact is practically realistic because typical design problems will involve a compact $\mathscr{X}$. Let H be the class of probability distributions on the Borel sets of $\mathscr{X}$.

Any $\eta \in$ H will be called a design *measure*.

For $\eta \in$ H define

$$M(\eta) = E(\tilde{x}\tilde{x}^T),$$

where $\tilde{x}$ is a random vector with distribution η and E denotes expected value. We shall refer to $M(\eta)$, a symmetric $k \times k$ matrix as *the information matrix of* η. Note that, because $\mathscr{X}$ is compact, $M(\eta)$ exists for all $\eta \in$ H. Let $\mathscr{M} = \{M(\eta): \eta \in \text{H}\}$.

Suppose that ϕ is a real-valued function defined on the $k \times k$ symmetric matrices and bounded above on $\mathscr{M}$, though possibly not below; that is we allow ϕ to take the value $-\infty$ on $\mathscr{M}$. For instance, if $\phi = \log \det$ and $M(\eta)$ is singular we would set $\phi\{M(\eta)\} = -\infty$.

Our problem is: *determine η_* to maximize $\phi\{M(\eta)\}$ over* H.

Any such η_* will be termed ϕ-optimal.

We now note some particular aspects of this problem.

3.2 *The set $\mathscr{M}$.*

3.2.1 Each element of $\mathscr{M}$ is a symmetric non-negative definite $k \times k$ matrix which can be represented by a point in $R^{(1/2)k(k+1)}$, the point with coordinates $\{m_{ij}; 1 \leq i \leq j \leq k\}$, when $M = (m_{ij})$.

3.2.2 The set $\mathscr{M}$ is convex. Indeed it is the closed convex hull of $\{xx^T : x \in \mathscr{X}\}$. Note that if η_x is the design measure putting probability 1 at x, then $M(\eta_x) = xx^T$.

3.2.3 By Carathéodory's Theorem each element of $\mathscr{M}$ can be expressed as a convex combination $\sum_{i=1}^{I} \lambda_i x_{(i)} x_{(i)}^T$, where $x_{(i)} \in \mathscr{X}$, $i = 1, \ldots, I$ and $I \leq \frac{1}{2}k(k+1) + 1$. By the same theorem, if M is a boundary point of $\mathscr{M}$, the inequality involving I can be strengthened to $I \leq \frac{1}{2}k(k+1)$: see Appendix 2.

From a practical point of view this remark is extremely important. For it means that if ϕ is maximal at M_* then M_* can always be expressed as $M(\eta_*)$, where η_* is a discrete design measure supported on at most $\frac{1}{2}k(k+1) + 1$ points; that is, there always exists a *discrete* η_* which solves the approximate theory problem. Our ultimate aim is to approximate η_* by a design measure corresponding to an N-observation design and this is clearly going to be easier if η_* is discrete.

3.3 *The criterion function ϕ.*

3.3.1 We have already mentioned one property of functions ϕ that arise from practical considerations. They are all increasing in the sense that if $M_1 - M_2$ is non-negative definite, then $\phi(M_1) \geq \phi(M_2)$. If, when $\phi(M_1)$ is finite and $M_1 - M_2$ is non-negative definite and non-zero, the inequality is strict, then we shall say that ϕ is strictly increasing. For example if $\phi(M) = \det M$, then ϕ is strictly increasing.

Note that if ϕ is strictly increasing its maximum must occur at a boundary point of $\mathscr{M}$. For if M is an interior point of $\mathscr{M}$ and $\mathscr{M}$ has dimension $\frac{1}{2}k(k+1)$ then there exists $a > 1$ such that $aM \in \mathscr{M}$ and then

$$\phi(aM) = \phi\{M + (a-1)M\} > \phi(M).$$

In this case there exists an optimal η_* supported on at most $\frac{1}{2}k(k+1)$ points: see Section 3.2.3.

If the dimension of $\mathscr{M}$ is less than $\frac{1}{2}k(k+1)$, ϕ may be maximal at a point in the relative interior of $\mathscr{M}$, but then every M can be expressed as a convex combination of fewer than $\frac{1}{2}k(k+1)$ points in the set $\{xx^T : x \in \mathscr{X}\}$.

3.3.2 Practical considerations provide a rather unexpected bonus. Criterion functions of practical interest turn out to be concave on $\mathcal{M}$, or at least they can be replaced by equivalent functions which are concave: see Kiefer (1974).

We shall assume from now on that ϕ is concave on $\mathcal{M}$; that is, for $0 \leq \lambda \leq 1$ and $M_1, M_2 \in \mathcal{M}$, $\phi\{\lambda M_1 + (1-\lambda)M_2\} \geq \lambda \phi(M_1) + (1-\lambda)\phi(M_2)$. If, for $0 < \lambda < 1$, and $\phi(M_1)$, $\phi(M_2)$ both finite, this inequality is strict then ϕ is strictly concave on $\mathcal{M}^+$, the subset of $\mathcal{M}$ where ϕ is finite.

When ϕ is strictly concave on $\mathcal{M}^+$, the maximizing M_* is unique, for were there distinct maximizing M's, M_{*1} and M_{*2} say, we would have $\phi\{\frac{1}{2}(M_{*1} + M_{*2})\}$ greater than the common value of $\phi(M_{*1})$ and $\phi(M_{*2})$—a contradiction.

3.4 *The solution set of design measures*

We note that if η_1 and η_2 are design measures, so, for $0 \leq \lambda \leq 1$ is $\lambda \eta_1 + (1-\lambda)\eta_2$. Hence the set of design measures is a convex set. Furthermore, by the definition of M,

$$M\{\lambda \eta_1 + (1-\lambda)\eta_2\} = \lambda M(\eta_1) + (1-\lambda)M(\eta_2).$$

Therefore if ϕ is concave on $\mathcal{M}$, $f(\eta) = \phi\{M(\eta)\}$ is a concave function on H. From this it readily follows that *the set H_* of ϕ-optimal design measures is a convex subset of* H. This is the most that we can say in general. Because of the possibility of different η's having the same M, strict concavity of ϕ does not imply strict concavity of f. So we can say nothing in general about the uniqueness of η_*. It transpires, however, that quite often in practice η_* is unique.

Finally we note that, for certain ϕ's, there may not exist a ϕ-optimal design measure; see Pukelsheim (1980); but this does not occur in the case of those ϕ's of primary interest to us.

3.5 *Directional derivatives*

These will play a basic role in our theory. There are two derivatives of interest.

3.5.1 *The Gâteaux derivative* of ϕ at M_1 in the direction of M_2 is:

$$G_\phi(M_1, M_2) = \lim_{\varepsilon \to 0^+} \frac{1}{\varepsilon} \{\phi(M_1 + \varepsilon M_2) - \phi(M_1)\}.$$

OPTIMAL DESIGN

As we shall see, differentiability of ϕ at M_1 eases the ϕ-optimal design measure problem considerably. Now differentiability of ϕ at M_1 implies that G_ϕ is linear in its second argument, that is,

$$G_\phi(M_1, \sum a_i M_i) = \sum a_i G_\phi(M_1, M_i),$$

for all real a_i: see Appendix 3 and Rockafellar (1970, p. 241). This fact is critical in our theory.

3.5.2 *The Fréchet derivative* of ϕ at M_1 in the direction of M_2 is:

$$F_\phi(M_1, M_2) = \lim_{\varepsilon \to 0^+} \frac{1}{\varepsilon} [\phi\{(1-\varepsilon)M_1 + \varepsilon M_2\} - \phi(M_1)],$$

and this derivative will serve our purposes better than the previous one.

We note:

(i) $M_1, M_2 \in \mathcal{M}$ implies $(1-\varepsilon)M_1 + \varepsilon M_2 \in \mathcal{M}$ and so $\phi\{(1-\varepsilon)M_1 + \varepsilon M_2\}$ is automatically defined.

(ii) Concavity of ϕ implies that

$$\frac{1}{\varepsilon}[\phi\{(1-\varepsilon)M_1 + \varepsilon M_2\} - \phi(M_1)]$$

is a non-increasing function of ε in $0 < \varepsilon \leqslant 1$; see Whittle (1971, p. 56). Hence when ϕ is concave, $F_\phi(M_1, M_2)$ exists if we allow the value $+\infty$.

(iii) By putting $\varepsilon = 1$ in (ii) we have the result that

$$F_\phi(M_1, M_2) \geqslant \phi(M_2) - \phi(M_1).$$

(iv) By definition, $F_\phi(M_1, M_2) = G_\phi(M_1, M_2 - M_1)$. Hence differentiability of ϕ implies that, if $\sum a_i = 1$,

$$F_\phi(M_1, \sum a_i M_i) = \sum a_i F_\phi(M_1, M_i).$$

If $\tilde{M}$ is a random matrix, ϕ is differentiable and E denotes expected value, we have, by this linearity,

$$E\{F_\phi(M_1, \tilde{M})\} = F_\phi\{M_1, E(\tilde{M})\}.$$

(v) By definition, for fixed M,

$$F_\phi(M, M) = 0.$$

(iv) Suppose that $\tilde{x}$ is a random k-vector with distribution η and, as usual, $M(\eta) = E(\tilde{x}\tilde{x}^T)$. Then by combining (iv) and (v) we have the result that if ϕ is differentiable at $M(\eta)$,

$$E[F_\phi\{M(\eta), \tilde{x}\tilde{x}^T\} = F_\phi\{M(\eta), E(\tilde{x}\tilde{x}^T)\} = F_\phi\{M(\eta), M(\eta)\} = 0.$$

With these preliminaries we are in a position to establish the main theoretical results which enable us to construct ϕ-optimal design measures.

3.6 **Theorem.** *When ϕ is concave on $\mathcal{M}$, η_* is ϕ-optimal if and only if $F_\phi\{M(\eta_*), M(\eta)\} \leq 0$ for all $\eta \in H$.*

Proof. Necessity: ϕ maximal at $M(\eta_*)$ implies

$$\phi\{(1-\varepsilon)M(\eta_*) + \varepsilon M(\eta)\} - \phi\{M(\eta_*)\} \leq 0$$

for all ε in $[0, 1]$ and all $\eta \in H$, since $(1-\varepsilon)M(\eta_*) + \varepsilon M(\eta) = M\{(1-\varepsilon)\eta_* + \varepsilon\eta\}$, and this in turn implies, from the definition of F_ϕ, that

$$F_\phi\{M(\eta_*), M(\eta)\} \leq 0, \quad \text{for all } \eta \in H.$$

Sufficiency: $F_\phi\{M(\eta_*), M(\eta)\} \leq 0$, for all $\eta \in H$, by 3.5.2(iii) implies $\phi\{M(\eta)\} - \phi\{M(\eta_*)\} \leq 0$, for all $\eta \in H$; that is, η_* is ϕ-optimal.

The proof of this theorem is really trivial and this is not surprising since what it says essentially is: we know that we are the top of a concave mountain when there is no direction in which we can look upward to another point on the mountain. However the theorem has little practical value since $\mathcal{M}$ is a set in $R^{(1/2)k(k+1)}$, and if k is at all large there are a great many directions in which we must look to determine whether or not the condition it contains is satisfied. Now if the mountain is smooth at the summit, that is, if it has no ridges running there it is not necessary to look around in *every* direction to verify that we are at the summit—we need only look towards extreme points of its convex base. This is the heuristic content of the following theorem.

3.7 **Theorem.** *If ϕ is concave on $\mathcal{M}$ and differentiable at $M(\eta_*)$ then η_* is ϕ-optimal if and only if $F_\phi\{M(\eta_*), xx^T\} \leq 0$ for all $x \in \mathcal{X}$.*

Proof. The necessity of the stated condition follows immediately from Theorem 3.6.

Sufficiency is proved as follows:
Any $M(\eta)$ can be expressed in the form

$$M(\eta) = \sum_{i=1}^{I} \lambda_i x_{(i)} x_{(i)}^T,$$

where $\sum \lambda_i = 1$ and each $\lambda_i > 0$. Then if ϕ is differentiable at $M(\eta_*)$

$$F_\phi\{M(\eta_*), M(\eta)\} = \sum_i \lambda_i F_\phi\{M(\eta_*), x_{(i)} x_{(i)}^T\}$$

see 3.5.2(iv). Hence $F_\phi\{M(\eta_*), xx^T\} \leq 0$ for all $x \in \mathscr{X}$ implies $F_\phi\{M(\eta_*), M(\eta)\} \leq 0$ for all $\eta \in H$ and the sufficiency now follows from Theorem 3.6.

This is the key theorem of ϕ-optimal linear regression design theory because it provides a condition which can often be verified quite easily in practice. The following example illustrates its use in a simple problem.

3.8 *Example. Quadratic regression with a single control variable*

Suppose that we have a single control variable u, real design space $U = [-1, 1]$, and that $E(\tilde{y}|u, \theta) = \theta_0 + \theta_1 u + \theta_2 u^2$. This means that the induced design space $\mathscr{X}$ is a subset of R^3. Indeed

$$\mathscr{X} = \left\{ x = \begin{pmatrix} 1 \\ u \\ u^2 \end{pmatrix}; \; -1 \leq u \leq 1 \right\},$$

which is an arc of a parabola in the plane $x_1 = 1$.

Suppose further than our criterion is D-optimality, that is,

$$\phi(M) = \log \det M, \quad \text{if } M \text{ is non-singular},$$
$$= -\infty, \quad \text{if } M \text{ is singular}.$$

Note that we take $\phi = \log \det$ rather than $\det$ to ensure concavity, and also that we shall be interested only in η's such that $M(\eta)$ is non-singular.

We start by calculating $F_\phi(M_1, M_2)$ for non-singular M_1. It is slightly easier, and this is usual, to do so via $G_\phi(M_1, M_2)$. We have

$$\log \det (M_1 + \varepsilon M_2) - \log \det M_1$$
$$= \log \det (I + \varepsilon M_2 M_1^{-1})$$
$$= \log \{1 + \varepsilon \operatorname{tr} (M_2 M_1^{-1})\} + O(\varepsilon^2)$$
$$= \varepsilon \operatorname{tr} (M_2 M_1^{-1}) + O(\varepsilon^2).$$

Hence
$$G_\phi(M_1, M_2) = \operatorname{tr} (M_2 M_1^{-1}).$$

This is linear in M_2 and Theorem 3.7 applies since only η for which $M(\eta)$ is non-singular can be optimal in this case.

Moreover
$$F_\phi(M_1, M_2) = G_\phi(M_1, M_2 - M_1) = \operatorname{tr} (M_2 M_1^{-1}) - 3,$$

and
$$F_\phi(M_1, xx^T) = x^T M_1^{-1} x - 3.$$

Hence according to Theorem 3.7, η_* is D-optimal if and only if
$$x^T \{M(\eta_*)\}^{-1} x \leqslant 3 \quad \text{for all } x \in \mathcal{X}.$$

We turn now to practical intuition. It is plausible that an $N = 3m$ observation design according to which we take m observations at each of the values $-1, 0, 1$ of u is at least a competitor to be a 'best' $3m$-observation design. The corresponding design measure η_0, say, puts probability $\frac{1}{3}$ at each of the vectors

$$\begin{pmatrix} 1 \\ -1 \\ 1 \end{pmatrix}, \quad \begin{pmatrix} 1 \\ 0 \\ 0 \end{pmatrix} \quad \text{and} \quad \begin{pmatrix} 1 \\ 1 \\ 1 \end{pmatrix}.$$

A trivial calculation shows that
$$M(\eta_0) = \frac{1}{3} \begin{pmatrix} 3 & 0 & 2 \\ 0 & 2 & 0 \\ 2 & 0 & 2 \end{pmatrix}.$$

With $x^T = (1 \ u \ u^2)$ we have
$$x^T \{M(\eta_0)\}^{-1} x = \tfrac{3}{4}\{4 - 6u^2(1-u^2)\}$$
$$\leqslant 3 \quad \text{for } -1 \leqslant u \leqslant 1,$$

that is for all $x \in \mathcal{X}$. Hence η_0 is indeed D-optimal.

OPTIMAL DESIGN

This particular use of Theorem 3.7 rests heavily on intuition to suggest a 'best' design. However as we shall see in the next chapter the theorem suggests algorithms for constructing ϕ-optimal designs in situations where intuition may fail.

In the meantime we consider some further general theory.

3.9 Theorem. *If ϕ is differentiable at all points of $\mathcal{M}^+$, the subset of $\mathcal{M}$ where $\phi(M) > -\infty$, and a ϕ-optimal measure exists, then η_* is ϕ-optimal if and only if*

$$\max_{x \in \mathcal{X}} F_\phi\{M(\eta_*), xx^T\} = \min_\eta \max_{x \in \mathcal{X}} F_\phi\{M(\eta), xx^T\}.$$

Here the minimum with respect to η is the minimum over $\{\eta : M(\eta) \in \mathcal{M}^+\}$.

Proof. We recall that if $\tilde{x}$ is a random vector with distribution η and ϕ is differentiable at $M(\eta)$, then

$$E[F_\phi\{M(\eta), \tilde{x}\tilde{x}^T\}] = 0;$$

see 3.5.2(vi). Hence for all η in $\{\eta : M(\eta) \in \mathcal{M}^+\}$ we have

$$\max_{x \in \mathcal{X}} F_\phi\{M(\eta), xx^T\} \geq 0.$$

However, according to Theorem 3.7, η_* is ϕ-optimal if and only if

$$\max_{x \in \mathcal{X}} F_\phi\{M(\eta_*), xx^T\} \leq 0,$$

and it follows that

$$\min_\eta \max_{x \in \mathcal{X}} F_\phi\{M(\eta), xx^T\} = 0,$$

with the minimum attained at a ϕ-optimal η_* when such an η_* exists. Hence the stated condition is necessary.

On the other hand if $\eta_\dagger$ satisfies the condition and a ϕ-optimal η exists, then $\max_{x \in \mathcal{X}} F_\phi\{M(\eta_\dagger), xx^T\} = 0$ and by Theorem 3.7, $\eta_\dagger$ is ϕ-optimal. Hence the condition is also sufficient.

3.10 Corollary. We note as a by-product of Theorem 3.9 that if η_* is ϕ-optimal and ϕ is differentiable at $M(\eta_*)$ then we have both

$$\max_{x \in \mathcal{X}} F_\phi\{M(\eta_*), xx^T\} = 0$$

and
$$E[F_\phi\{M(\eta_*), \tilde{x}\tilde{x}^T\}] = 0,$$
where $\tilde{x}$ is a random vector with distribution η_*. This can happen only if $F_\phi\{M(\eta_*), \tilde{x}\tilde{x}^T\} = 0$ with η_* probability 1. If η_* is discrete with finite support $x_{(1)}, \ldots, x_{(I)}$ then
$$F_\phi\{M(\eta_*), x_{(i)}x_{(i)}^T\} = 0, \qquad i = 1, 2, \ldots, I.$$

3.11 From a practical point of view Theorem 3.9 is not nearly as fundamental as Theorem 3.7, since it does not provide such a readily verifiable condition for ϕ-optimality. Its interest is mainly confined to the case where $\phi = \log \det$. In this case, as we have seen in Example 3.8,
$$F_\phi\{M, xx^T\} = x^T M^{-1} x - k.$$
Now by definition a design measure η_* is G-optimal if
$$\max_{x \in \mathcal{X}} x^T \{M(\eta_*)\}^{-1} x = \min_\eta \max_{x \in \mathcal{X}} x^T \{M(\eta)\}^{-1} x;$$
see Section 2.2.3. Therefore by taking $\phi = \log \det$ in Theorem 3.9 we establish the equivalence *for design measures*, of D- and G-optimality. Note that this equivalence does not necessarily hold for N-observation designs.

The particular versions of Theorems 3.7 and 3.9 obtained by taking $\phi = \log \det$ provide the essential content of the celebrated Equivalence Theorem of Kiefer & Wolfowitz (1960).

3.12 Theorem 3.7 provides us with a powerful theoretical tool. However do we have to worry in practice about the existence of η such that $G_\phi\{M(\eta), M(\eta')\}$ is non-linear in its second argument, so that the sufficiency of the condition stated in Theorem 3.7 no longer obtains? Unfortunately the answer is yes. The following example shows why.

Suppose that $x^T = (x_1, x_2)$; $\mathcal{X}$ is the quadrilateral with vertices $(0, 0)$, $(1, 0)$, $(4, 1)$, $(4, 2)$; $E(\tilde{y}|x, \theta) = \theta_1 x_1 + \theta_2 x_2$; and we wish to minimize the variance of the least-squares estimator of θ_1, so that θ_2 may be regarded as a nuisance parameter. All N-observation designs with non-singular information matrices admit estimation of θ_1; so do N-observation designs where all observations are taken at points in the interval $(0, 1]$ on the x_1-axis; no other N-observation design does. If an N-observation design $\mathbf{x}$ has non-singular information matrix $M(\mathbf{x}) = (m_{ij}(\mathbf{x}))$, then the variance of the least-squares estimator of θ_1 obtained from it is proportional to $\{m_{11}(\mathbf{x}) - m_{12}^2(\mathbf{x})/m_{22}(\mathbf{x})\}^{-1}$; if it has a singular information matrix with $m_{22}(\mathbf{x}) = m_{12}(\mathbf{x}) = 0$ and $m_{11}(\mathbf{x}) \neq 0$, this variance is proportional to $m_{11}^{-1}(\mathbf{x})$; otherwise θ_1 is not estimable.

Hence for design measures an appropriate definition of the criterion function ϕ to be maximized is

$$\phi\{M(\eta)\} = \log\{m_{11}(\eta) - m_{12}^2(\eta)/m_{22}(\eta)\},$$
$$\text{when } M(\eta) \text{ is non-singular},$$
$$= \log m_{11}(\eta),$$
$$\text{when } m_{11}(\eta) \neq 0, m_{12}(\eta) = m_{22}(\eta) = 0,$$
$$= -\infty, \quad \text{otherwise}.$$

Here we introduce the logarithmic function for convenience in subsequent calculation.

Now consider the design measure η_0 which puts probability 1 at $(1, 0)$. This is a candidate to be a 'best' design for estimating θ_1, that is, a ϕ-optimal design. We have

$$M(\eta_0) = \begin{pmatrix} 1 & 0 \\ 0 & 0 \end{pmatrix},$$

and an elementary calculation shows that

$$F_\phi\{M(\eta_0), M(\eta)\} = m_{11}(\eta) - m_{12}^2(\eta)/m_{22}(\eta) - 1, \quad \text{if } m_{22}(\eta) \neq 0$$
$$= m_{11}(\eta) - 1, \quad \text{if } m_{12}(\eta) = m_{22}(\eta) = 0.$$

This is not appropriately linear in its second argument and the implications of this non-linearity for the use of Theorem 3.7 are easily seen. We have, in particular,

$$F_\phi\{M(\eta_0), xx^T\} = -1, \quad \text{if } x_2 \neq 0,$$
$$= x_1^2 - 1, \quad \text{if } x_2 = 0.$$

Thus for all $x \in \mathscr{X}$, $F_\phi\{M(\eta_0), xx^T\} \leq 0$. However we cannot conclude that η_0 is ϕ-optimal. Indeed it is not, since for the design measure η_1 which puts probability $\frac{1}{2}$ at each of $(4, 1)$ and $(4, 2)$, it is readily shown that

$$F_\phi\{M(\eta_0), M(\eta_1)\} = \tfrac{3}{5},$$

and η_0 fails to satisfy the necessary and sufficient condition of Theorem 3.6, which applies whether or not ϕ is differentiable at $M(\eta_0)$.

We note in passing that ϕ is differentiable at non-singular M, that for such M,

$$F_\phi(M, xx^T) = \frac{(x_1 - m_{12}m_{22}^{-1}x_2)^2}{m_{11} - m_{12}^2 m_{22}^{-1}} - 1,$$

and that the design measure putting probability $\frac{2}{3}$ at $(4, 1)$, $\frac{1}{3}$ at $(4, 2)$ is ϕ-optimal.

This is a particular example of D_s-optimality, which we shall discuss in general later. Typically we have to face the problem of non-differentiability of ϕ when this function is finite at singular M, though it does occur in certain other cases, notably in connection with E-optimality; see Section 2.2.4.

3.13 Optimal design measures with singular information matrices

We shall now consider in some generality the problem illustrated in Section 3.12 arising from the possibility that a design measure with singular information matrix may be optimal.

Suppose that we are interested in certain linear combinations of the unknown parameters, say the vector $A^T\theta$, where A^T is an $s \times k$ matrix of rank $s < k$. Since $s < k$, certain N-observation designs with singular information matrices allow estimation of $A^T\theta$. Accordingly in the approximate theory we have to take account of this. Now by analogy with general results in least-squares theory we can say that a design measure with information matrix M allows estimation of $A^T\theta$ if $Mz = 0$ implies $A^Tz = 0$, or equivalently if $A = MY$, for some matrix Y. Let $\mathcal{M}_A$ be the subset of $\mathcal{M}$ consisting of matrices with this property; $\mathcal{M}_A$ contains all the non-singular matrices in $\mathcal{M}$ and also some singular ones. Again by analogy with results in least-squares theory, the variance matrix of the least-squares estimator of $A^T\theta$ arising from a design measure with information matrix $M \in \mathcal{M}_A$ is proportional to $A^T M^- A$, where M^- is *any* generalized inverse (g-inverse) of M, that is any matrix such that $MM^-M = M$; see, for instance, Searle (1971, p. 184). Note that $A^T M^- A$ does not depend on which g-inverse is used; also that $A^T M^- A$ is a positive-definite $s \times s$ matrix. Interest in $A^T\theta$ implies concern about making

$A^T M^- A$ small in some sense. For instance we might wish to minimize $\log \det (A^T M^- A)$ or $\text{tr}(A^T M^- A)$; or equivalently to maximize $-\log \det (A^T M^- A)$ or $-\text{tr}(A^T M^- A)$. The first of these criteria is of course just the D_A-optimal one introduced in Section 2.2.2 and the second is the linear criterion of Section 2.2.5, extended here to allow for singular $M \in \mathcal{M}_A$.

More generally let us suppose that we wish to maximize a function $\phi(M)$ defined by

$$\phi(M) = \psi(A^T M^- A), \quad \text{if } M \in \mathcal{M}_A$$
$$= -\infty, \quad \text{otherwise}.$$

Here ψ is a function defined and finite on the positive-definite $s \times s$ matrices. We shall assume that ϕ is concave on $\mathcal{M}_A$, as is the case when $\psi = -\log \det$ or $\psi = -\text{tr}$; also that ϕ is differentiable at non-singular M and non-differentiable at singular M, again something which is true for these particular ϕ.

If M_* is a non-singular matrix in $\mathcal{M}_A$ then we can appeal to Theorem 3.7 and say that M_* is the information matrix of a ϕ-optimal measure if and only if $F_\phi(M_*, xx^T) \leq 0$ for all $x \in \mathcal{X}$. However suppose that $M_* \in \mathcal{M}_A$ and has rank $r < k$. To emphasize this we shall write M_r rather than M_*. Because of the non-differentiability of ϕ at M_r we can no longer appeal to Theorem 3.7, just as in the example in Section 3.12, and we are thrown back on the practically less useful Theorem 3.6 to verify the optimality, or otherwise, of M_r. We require a more readily verifiable condition in the case of singular M_r than that stated in Theorem 3.6, and it is probably true to say that, at the time of writing, this is the main outstanding problem in the approximate theory of optimal *linear* regression design.

Silvey (1978) has made some progress in the solution of this problem and further progress has been made by Pukelsheim (1980), and we now consider the present author's contribution. Note that since ψ is defined on the positive definite $s \times s$ matrices, ϕ is well defined on the class $\mathcal{L}_A$ of non-negative definite $k \times k$ matrices L such that $A = LY$ for some Y. Of course $\mathcal{L}_A$ includes all the positive definite $k \times k$ matrices as well as $\mathcal{M}_A$. Now, given M_r of rank $r < k$, there exist matrices H of order $k \times (k - r)$ and rank $k - r$, such that $M_r + HH^T$ is non-singular; any H whose columns together with those of M_r, span R^k, has this property.

Let $\mathscr{H}(M_r)$ be the class of all such matrices. Then we have the following.

3.13.1 Theorem. *Let ϕ, M_r and $\mathscr{H}(M_r)$ be defined as above. A sufficient condition that ϕ is maximal on $\mathscr{M}$ at M_r is that there exists an $H \in \mathscr{H}(M_r)$ such that $F_\phi(M_r + HH^T, xx^T) \leq 0$ for all $x \in \mathscr{X}$.*

Proof. Since $M_r + HH^T$ is non-singular, ϕ is differentiable there. Hence, as in Theorem 3.7 the stated condition implies $\phi(M_r + HH^T) \geq \phi(M)$ for all $M \in \mathscr{M}$. However $(M_r + HH^T)^{-1}$ is a g-inverse of M_r: see Searle (1971, p. 22), and so $\phi(M_r + HH^T) = \phi(M_r)$. Consequently $\phi(M_r) \geq \phi(M)$ for all $M \in \mathscr{M}$.

This proof of the sufficiency of the condition is remarkably simple. A much more difficult question is whether it is also necessary. It is possible to show that it is necessary when $\psi = -\log \det$ and $\psi = -\text{tr}$; and it is probably necessary for many more ψ. However this remains an open question. Also open are the questions of how to determine an H when one exists and, for cases where the condition is necessary, how to show that no H exists satisfying the condition: see Silvey (1978).

Pukelsheim (1980) uses duality theory to obtain a necessary and sufficient condition for a design measure with singular information matrix to be ϕ-optimal, for a fairly wide class of functions ϕ—the ϕ_p functions introduced by Kiefer (1974). Once again, though, a practical method of verifying this condition has yet to be devised.

We shall return to this difficulty when discussing particular criteria in a subsequent chapter. But first we shall discuss algorithms that make the general theory we have considered practically useful.

4
ALGORITHMS

4.1 Introduction

Theorem 3.7 was used in Example 3.8 to verify that an intuitively sensible design measure was in fact D-optimal. However to exploit its full potential we require more than this; we need algorithms that enable us to construct ϕ-optimal design measures. Now we know that there always exists an optimal measure with finite support. Suppose that we know *a priori* a finite collection of points $x_{(1)}, \ldots, x_{(T)}$, say, among which the support points of a ϕ-optimal measure lie. Then the only problem is to find the appropriate probabilities at these points. This is a standard extremum problem; we have a concave function

$$f(\eta_1, \ldots, \eta_T) = \phi(\sum \eta_i x_{(i)} x_{(i)}^T)$$

that we wish to maximize subject to $\eta_i \geq 0$, $\sum \eta_i = 1$; and it can be tackled by standard numerical methods; see Wu (1978); also Silvey, Titterington & Torsney (1978) for a purpose-built algorithm.

In fact it is possible on occasion to use the theory that we have developed to identify such a finite collection of points, even when the design space is infinite. The following simple example, due to Wynn (1970) illustrates the point. Suppose that $x^T = (1, x_1, x_2)$; $\mathcal{X}$ is the quadrilateral in R^3 with vertices $x_{(1)}^T = (1, 2, 2)$, $x_{(2)}^T = (1, -1, 1)$, $x_{(3)}^T = (1, 1, -1)$, $x_{(4)}^T = (1, -1, -1)$; $E(\tilde{y}|x, \theta) = \theta_0 + \theta_1 x_1 + \theta_2 x_2$; and our criterion is D-optimality: $\phi(M) = \log \det M$. Theorem 3.7 tells us that η_* is D-optimal if and only if $x^T \{M(\eta_*)\}^{-1} x \leq 3$ for all $x \in \mathcal{X}$, and Corollary 3.10 says that $x^T \{M(\eta_*)\}^{-1} x$ takes its maximum value of 3 at the support points of η_*. Now $x^T \{M(\eta_*)\}^{-1} x$ is a strictly convex function of x and so its maximum on the convex set $\mathcal{X}$ can occur only at extreme points of $\mathcal{X}$. Hence the support points of η_* must be contained within this set of extreme points, that is, the vertices $x_{(1)}, x_{(2)},$

$x_{(3)}$, $x_{(4)}$. If η is the design measure that puts probability η_i at $x_{(i)}$, $i = 1, \ldots, 4$, then

$$M(\eta) = \sum \eta_i x_{(i)} x_{(i)}^T$$

and we have to choose $\eta_1, \ldots, \eta_4$ to maximize det $M(\eta)$.

This reduced problem has the mild complication that one of the optimum η_i's may be zero; not more than one can be, since otherwise $M(\eta)$ would be singular. If we happened to know that all four vertices were support points of η_* then all that would be necessary would be to solve for η the equations

$$x_{(i)}^T \{M(\eta)\}^{-1} x_{(i)} = 3, \qquad i = 1, \ldots, 4,$$

to find η_*. In fact if we try this in this example we do obtain a feasible solution since all four vertices are indeed support points of the optimal η_*.

4.2 Design measure algorithms

The problem of constructing ϕ-optimal design measures is not so standard if we cannot initially identify a finite set which includes the support points of an optimal measure, and it is this fact that has led to the development of special algorithms for this purpose. These are hill-climbing algorithms which exploit the fact that the extreme points of the convex set $\mathcal{M}$ are included in the set $\{xx^T : x \in \mathcal{X}\}$. The basic idea underlying the earliest of them is simple; see Wynn (1970) and Fedorov (1972). Suppose that we have a design measure η_n with information matrix M_n and that ϕ is differentiable at M_n. We determine $x_{(n+1)}$ to maximize $F_\phi(M_n, xx^T)$ over $\mathcal{X}$. If η_n is not ϕ-optimal, then by Theorem 3.7 this maximum value is greater than zero. Hence we can increase ϕ by moving from M_n in the direction of $x_{(n+1)} x_{(n+1)}^T$. This we do by putting more weight at $x_{(n+1)}$ than η_n puts there. Accordingly we choose α_{n+1} in the interval $(0, 1)$ and define η_{n+1} by

$$\eta_{n+1} = (1 - \alpha_{n+1})\eta_n + \alpha_{n+1} \eta_{x_{(n+1)}},$$

where $\eta_{x_{(n+1)}}$ is the measure that puts probability 1 at $x_{(n+1)}$. Note that then M_{n+1}, the information matrix of η_{n+1} is given by

$$M_{n+1} = (1 - \alpha_{n+1}) M_n + \alpha_{n+1} x_{(n+1)} x_{(n+1)}^T,$$

so that by putting more weight at $x_{(n+1)}$ we are indeed moving from M_n in the direction of $x_{(n+1)}x_{(n+1)}^T$.

This leaves open the choice of *step-length*, α_{n+1}. Not all choices will ensure that $\phi(M_{n+1}) > \phi(M_n)$. Certainly by taking α_{n+1} small enough we can ensure this inequality, since $F_\phi(M_n, x_{(n+1)}x_{(n+1)}^T) > 0$. But if we take too large a step-length we may go over the crest in the direction of $x_{(n+1)}x_{(n+1)}^T$ and far enough down thereafter to make $\phi(M_{n+1}) < \phi(M_n)$.

One possibility is to choose α_{n+1} to give maximal increase in ϕ and, assuming differentiability of ϕ between M_n and $x_{(n+1)}x_{(n+1)}^T$, the α_{n+1} which does so is that determined by $F_\phi(M_{n+1}, x_{(n+1)}x_{(n+1)}^T) = 0$, the step-length that takes us exactly to the crest in the chosen direction. An iterative algorithm whose typical iteration is as described, with this optimum step-length used, we shall refer to as a V-algorithm—V for V.V. Fedorov (1972) who discusses it for various specific ϕ.

Another way of choosing step-lengths for this type of iterative algorithm is to fix a sequence (α_n) of step-lengths in advance, without worrying about whether ϕ increases on each iteration. Commonsense suggests that if we do so, in order to have any hope of getting close to an optimal M_* we must require $\alpha_n \to 0$ as $n \to \infty$; otherwise we might simply go back and forwards over the summit; and also $\sum \alpha_n$ divergent since otherwise we might stutter to a stop on the hillside. An iterative algorithm using such a sequence (α_n) will be referred to as a W-algorithm—W for Wynn (1970), who considered a special case of it; see also Fedorov (1972).

Before further general discussion of these algorithms we shall consider particular examples.

4.3 *The V-algorithm for D-optimality*

We consider the case where

$$\phi(M) = \log \det M, \quad M \text{ non-singular}$$
$$= -\infty, \quad \text{otherwise}.$$

Then for M where $\phi(M)$ is finite, we have

$$F_\phi(M, xx^T) = x^T M^{-1} x - k.$$

ALGORITHMS

We start with a design measure η_0 having non-singular information matrix M_0. Note the importance of having M_0 non-singular. Typically η_0 will put probability $1/k$ at each of k linearly independent vectors in $\mathscr{X}$. Then at stage n in the V-algorithm we will have a design measure η_n with non-singular matrix M_n. Note that when M_0 is non-singular, M_n is necessarily non-singular because there is no possibility of any step-length being unity. Therefore M_n = a positive constant $\times M_0$ + a non-negative definite matrix, and so it is non-singular.

We shall write

$$d_n = x^T M_n^{-1} x$$

and

$$\bar{d}_n = \max_{x \in \mathscr{X}} x^T M_n^{-1} x = x_{(n+1)}^T M_n^{-1} x_{(n+1)},$$

by the definition of $x_{(n+1)}$. Then

$$M_{n+1} = (1 - \alpha_{n+1}) M_n + \alpha_{n+1} x_{(n+1)} x_{(n+1)}^T,$$

and by a standard matrix result,

$$M_{n+1}^{-1} = \{(1 - \alpha_{n+1}) M_n\}^{-1} - \frac{\alpha_{n+1} M_n^{-1} x_{(n+1)} x_{(n+1)}^T M_n^{-1}}{(1 - \alpha_{n+1})^2 \{1 + \alpha_{n+1} \bar{d}_n/(1 - \alpha_{n+1})\}}$$

Hence

$$F_\phi(M_{n+1}, x_{(n+1)} x_{(n+1)}^T) = x_{(n+1)}^T M_{n+1}^{-1} x_{(n+1)} - k$$
$$= \frac{\bar{d}_n}{1 + \alpha_{n+1}(\bar{d}_n - 1)} - k,$$

after simplication.

The optimum step-length, α_{n+1}, given by

$$F_\phi(M_{n+1}, x_{(n+1)} x_{(n+1)}^T) = 0,$$

is therefore determined by

$$\frac{\bar{d}_n}{1 + \alpha_{n+1}(\bar{d}_n - 1)} = k,$$

and so

$$\alpha_{n+1} = \frac{\bar{d}_n - k}{k(\bar{d}_n - 1)}.$$

OPTIMAL DESIGN

The V-algorithm is therefore very easy to apply, in this case. Our initial iteration involves calculating M_0^{-1}; determining $x_{(1)}$ to maximize $x^T M_0^{-1} x$ over $\mathscr{X}$, the maximum value being $\bar{d}_0$; calculating $\alpha_1 = (\bar{d}_0 - k)/\{k(\bar{d}_0 - 1)\}$; and then computing $M_1 = (1 - \alpha_1)M_0 + \alpha_1 x_{(1)} x_{(1)}^T$. Computationally the only non-trivial exercise in this iteration is that of determining $x_{(1)}$. This remark applies equally to subsequent iterations; indeed if we use the standard matrix result referred to above we do not even have a matrix inversion in iterations subsequent to the first.

Of course we gain much here by obtaining an explicit, simple expression for the optimum step length, and this may not be possible for ϕ's other than log det.

It is not difficult to prove that in the D-optimal case the sequence (M_n) generated by the V-algorithm converges to M_*, the unique M where ϕ is maximal; see Fedorov (1972, p. 102). Note that M_* is unique because log det is strictly concave; see Section 3.3.2. If the optimal η_* happens also to be unique then of course the generated sequence (η_n) converges to it.

4.4 *An example of the use of the W-algorithm*

In order to illustrate some of the features of the W-algorithm we consider its application in a very simple situation where an analytic solution is possible.

Suppose that we have one control variable u, real design space $\mathscr{U} = [0, 1]$ and that $E(\tilde{y}|u, \theta) = \theta_0 + \theta_1 u$—simple linear regression, in fact. We wish to design in order to predict the expected value of $\tilde{y}$ at the value a of u, where $a > 1$.

The induced design space $\mathscr{X}$ is the line segment joining the points $(1, 0)$ and $(1, 1)$ in the plane. Let $c^T = (1 \quad a)$. The appropriate criterion function ϕ is

$$\phi(M) = -c^T M^{-1} c, \quad \text{if } M \text{ is non-singular}$$
$$= -\infty, \quad \text{otherwise},$$

since no design measure with singular information matrix allows estimation of $c^T \theta$. An elementary calculation shows that if $\phi(M_1)$ is finite,

$$G_\phi(M_1, M_2) = c^T M_1^{-1} M_2 M_1^{-1} c,$$

which is linear in M_2. Therefore Theorem 3.7 applies.

We have

$$F_\phi(M, xx^T) = (c^T M^{-1} x)^2 - c^T M^{-1} c.$$

For $M \in \mathcal{M}$, this is strictly convex on $\mathcal{X}$ and by the argument in Section 4.1 the only possible support points of an optimal measure are $x_{(1)}^T = (1 \ \ 0)$ and $x_{(2)}^T = (1 \ \ 1)$; these must both be support points since we require at least two for a non-singular design. If we now consider the design measure η that puts probability p at $x_{(2)}$, $1-p$ at $x_{(1)}$ then it is very easy to show that $\phi\{M(\eta)\}$ is maximal when $p = a/(2a-1)$. Hence the unique optimal η_* puts this probability at $x_{(2)}$. This is the analytic solution to the problem.

Imagine that we do not know this. Suppose that $a = \frac{5}{4}$ and let us apply the W-algorithm, taking as starting design measure η_0 that which puts probability $\frac{1}{2}$ at each of $x_{(1)}$ and $x_{(2)}$, so that

$$M_0 = \frac{1}{2}\begin{pmatrix} 2 & 1 \\ 1 & 1 \end{pmatrix},$$

and using the sequence $\frac{1}{3}, \frac{1}{4}, \ldots$ of step-lengths. This choice of step-lengths is natural if we think about starting with a 2-observation design, adding another point to give a 3-observation design, and so on; it yields a sequence of design measures corresponding to these designs and the design measure corresponding to an N-observation design is easily written down—the probability associated by it with a point is simply the number of times the point occurs in the N-observation design, divided by N. It was this idea that Wynn (1970) introduced.

An easy calculation shows that $F_\phi(M_0, xx^T)$ is maximized at $x_{(2)}$, and so η_1 puts probability $\frac{1}{3}$ at $x_{(1)}$, $\frac{2}{3}$ at $x_{(2)}$; and

$$M_1 = \frac{1}{3}\begin{pmatrix} 3 & 2 \\ 2 & 2 \end{pmatrix}.$$

Proceeding in this way we find that the first four iterations all add $x_{(2)}$ to the starting design, so that η_4 puts probability $\frac{5}{6}$ at $x_{(2)}$, $\frac{1}{6}$ at $x_{(1)}$. This of course is optimal and the algorithm has yielded an optimal design in four iterations.

4.5 *General comments*

We now make some general comments on algorithms in the light of these examples.

(i) Example 4.4 presents the W-algorithm in the most favourable possible light. Our choice of starting design was judicious since it included only optimal design support points. It is clear that had η_0 put positive probability at any x other than $x_{(1)}$ or $x_{(2)}$, the algorithm would not have yielded an optimal measure in a finite number of steps. In general speed of convergence will depend crucially on η_0. Vuchkov (1977) proposes, in a particular context, an exchange algorithm for a good choice of η_0, but whether a general version of such an algorithm would be advantageous depends on the amount of computation it would involve.

(ii) It is also clear that the W-algorithm, with the sequence (α_n) chosen as in Section 4.4, can only yield an optimal measure in a finite number of steps if the probabilities assigned by an optimal measure are rational numbers. In this respect too the example chosen was favourable to it.

(iii) Again referring to Example 4.4, whatever choice of η_0 had been made neither the V- nor the W-algorithm would ever have introduced an x other than $x_{(1)}$ or $x_{(2)}$. This is because of the convexity on $\mathscr{X}$ of $F_\phi(M, xx^T)$, mentioned above. Now it does often happen that $F_\phi(M, xx^T)$ is a convex function of x on convex subsets of R^k. However $\mathscr{X}$ is not always a convex set and even if it is, not all of its extreme points are necessarily optimal design support points. So we cannot expect in general that either the V- or the W-algorithms will introduce only optimal design support points. Eventually both algorithms begin to cycle on a finite set of points which are support points for an optimal design measure. In some cases this cycling starts quite early; in others it is slower; Tsay (1976) and Atwood (1976).

(iv) Both these early algorithms have been criticized on the grounds of their slow convergence and various suggestions have been advanced for speeding this; see Atwood (1973), Silvey & Titterington (1973) and Wu (1978). We emphasize again that what is really critical is the identification of optimal design measure support points. When these have been identified, the numerical analysis problem is a standard one. No modification of the early algorithms has altered, in any essential way, the basic notion of identifying optimal support points by introducing as a new point at stage n, an x which maximizes $F_\phi(M_n, xx^T)$.

4.6 Convergence considerations

Much attention has been paid to proofs of convergence to an optimum η_* of the V-, W- and other algorithms; Fedorov (1972), Wynn (1970, 1972), Atwood (1973) and others. Some of these proofs are certainly not easy, particularly these for the non-monotonic W-algorithm. However what is important about an algorithm is not whether it converges, but whether it is effective in the sense that it guarantees arbitrary close approach to the optimum; and how fast this approach is. Speed of approach, in the sense of computing time required is usually best dealt with empirically; Wu (1978). Proof of arbitrary close approach is, not surprisingly, much easier than proof of convergence in the non-monotonic case, and we illustrate this by considering the W-algorithm with a sequence (α_n) of step-lengths such that $0 < \alpha_n < 1$, $\alpha_n \to 0$ as $n \to \infty$ and $\sum \alpha_n$ diverges.

We note first that in practice we must have a stopping rule for any iterative algorithm and the natural stopping rule for those we have considered is: continue iteration until $\max_{x \in \mathscr{X}} F_\phi(M_n, xx^T) < \delta$, where δ is a pre-assigned small positive number. Suppose that we use this stopping rule. *Assume that ϕ is differentiable at M_n*. We now show that, with this assumption, the rule ensures that, when we stop,

$$\phi(M_n) \geqslant \phi(M_*) - \delta,$$

where M_* is the information matrix of a ϕ-optimal design measure.

By Section 3.2.3, M_* can be expressed as a finite convex combination $\sum \lambda_i x_{(i)} x_{(i)}^T$, and therefore, since ϕ is differentiable at M_n,

$$F_\phi(M_n, M_*) = \sum \lambda_i F_\phi(M_n, x_{(i)} x_{(i)}^T) \leqslant \max_{x \in \mathscr{X}} F_\phi(M_n, xx^T).$$

Now by Section 3.5.2(iii),

$$F_\phi(M_n, M_*) \geqslant \phi(M_*) - \phi(M_n),$$

and these inequalities together ensure that

$$\phi(M_*) - \phi(M_n) \leqslant \max_{x \in \mathscr{X}} F_\phi(M_n, xx^T) < \delta,$$

at the stopping stage.

OPTIMAL DESIGN

Do we necessarily reach this stopping stage? Consider the W-algorithm with a sequence of step-lengths having the above properties. Again suppose that ϕ *is a differentiable at all M_n* and that $\max_{x \in \mathscr{X}} F_\phi(M_n, xx^T) \geq \delta > 0$ for all n. We have—see Appendix 3:

$$\begin{aligned}\phi(M_n) - \phi(M_{n-1}) &= F_\phi(M_{n-1}, M_n) + o\|M_n - M_{n-1}\| \\ &= F_\phi\{M_{n-1}, (1-\alpha_n)M_{n-1} + \alpha_n x_{(n)} x_{(n)}^T\} + o(\alpha_n) \\ &= \alpha_n F_\phi(M_{n-1}, x_{(n)} x_{(n)}^T) + o(\alpha_n) \\ &= \alpha_n \max_{x \in \mathscr{X}} F_\phi(M_{n-1}, xx^T) + r_n,\end{aligned}$$

where $r_n = o(\alpha_n)$.

Since by assumption, for all n, $\max_{x \in \mathscr{X}} F_\phi(M_n, xx^T) \geq \delta$, we therefore have

$$\phi(M_n) - \phi(M_0) \geq \delta \sum_{i=1}^n \alpha_i + \sum_{i=1}^n r_i,$$

and the right-hand side tends to infinity as $n \to \infty$, since $\sum \alpha_i$ diverges, $r_n = o(\alpha_n)$, and consequently $\sum r_i / \sum \alpha_i \to 0$. Hence $\phi(M_n) \to \infty$ as $n \to \infty$. However $\phi(M_n)$ is bounded above, and this contradiction establishes that $\max_{x \in \mathscr{X}} F_\phi(M_n, xx^T) \not\geq \delta$, for all n. Since δ is arbitrary, this shows that the W-algorithm does indeed take us arbitrarily close to $\phi(M_*)$, *so long as ϕ is differentiable at all M_n*.

Note that we do not require here that ϕ be everywhere differentiable, nor that it be differentiable at M_*. Now, as we have previously noted, it is often the case that ϕ is non-differentiable at singular M, but differentiable at non-singular M. In this case we can ensure that ϕ is differentiable at M_n, for all n, simply by starting with a non-singular M_0. The fact that each $\alpha_n < 1$ then ensures, as in Section 4.3, that M_n is non-singular for all n.

4.7 Review

It seems appropriate at this point to review briefly the general approximate theory that we have considered to date and to discuss its implications for the N-observation design problem.

If ϕ, the concave criterion function, is differentiable at all points of $\mathscr{M}^+$, the subset of $\mathscr{M}$ where $\phi(M) > -\infty$, the situation is entirely satisfactory. We have effective algorithms for computing an optimal η_* and a relatively easy check on whether a given design measure is ϕ-optimal; see Theorem 3.7.

If ϕ is not everywhere differentiable on $\mathcal{M}^+$, the situation is less clear-cut. Algorithms are still available. If an algorithm, at stage n, reaches a point M_n where ϕ is differentiable and $F_\phi(M_n, xx^T) \leq 0$ for all $x \in \mathcal{X}$, then we are in the happy position of the previous paragraph. Again if an algorithm, having avoided points where ϕ is non-differentiable reaches a point M_n where ϕ is differentiable and which is such that $\max_{x \in \mathcal{X}} F_\phi(M_n, xx^T)$ is small then we know that η_n is close to an optimum η_*, and from a practical point of view this is all that really matters.

If, on the other hand, an algorithm reaches an M_n where ϕ is non-differentiable and which is such that $F_\phi(M_n, xx^T) \leq 0$ for all $x \in \mathcal{X}$, we have no ready means of determining whether or not η_n is ϕ-optimal. One practical possibility suggests itself; see Atwood (1969). If we can find an η, close to η_n, such that ϕ is differentiable at $M(\eta)$ and $\max_{x \in \mathcal{X}} F_\phi\{M(\eta), xx^T\}$ is small, then η_n is close to an η which is close an optimum η_* and so it is good enough for practical purposes. However if we cannot find such an η we cannot be sure about the status of η_n, without appealing to Theorem 3.13.1 in cases where the condition stated is necessary as well as sufficient; and even then we do not as yet have a ready means of verifying whether or not this condition is satisfied by a given η_n. It would be a mistake, however, to over-emphasize the practical import of this remaining weakness in the general theory. In the majority of practical problems, ϕ will be differentiable at an optimal M_* and algorithms will lead us there.

4.8 *N-observation designs*

From a practical point of view the object of considering approximate theory was to help with the more intractable N-observation design problem. Suppose that we have found an optimal design measure η_* supported on the finite set of points $x_{(1)}, \ldots, x_{(I)}$ and putting probabilities $\eta_{*1}, \ldots, \eta_{*I}$ at them respectively. If $N \gg I$ we can approximate each η_{*i} by a number $\eta_i = r_i/N$, where each r_i is an integer and $\sum r_i = N$. Then the design measure η is close to η_* and it is of course the design measure corresponding to the N-observation design according to which we take r_i observations at $x_{(i)}$, $i = 1,\ldots, I$. This N-observation design must then be at least close to a best one. It is not difficult to devise systematic ways of approximating η_* in this way; see Fedorov (1972, p. 157).

However if N is not large relative to I then often any approximating η obtained in this way will not be all that close to η_*. One point that does emerge from the approximate theory is worth noting, though. Suppose that we consider two design measures η_* and η, both with finite support. Let the union of their support points be $\{x_{(1)},\ldots, x_{(J)}\}$, let η put probabilities $\eta_1,\ldots,\eta_J$ at these points respectively, and similarly for η_*. Note that some of the η_i's and η_{*i}'s may be zero. Write

$$f(\eta) = f(\eta_1,\ldots,\eta_J) = \phi\left(\sum_{j=1}^{J} \eta_j x_{(j)} x_{(j)}^T\right) = \phi\{M(\eta)\}.$$

Then if we assume that f is twice differentiable, it is not difficult to show that

$$f(\eta) = f(\eta_*) + F_\phi\{M(\eta_*), M(\eta)\} + O(\|\eta_* - \eta\|^2),$$

where $\|\ \|$ denotes Euclidean distance.

Suppose now that η_* is ϕ-optimal. If the support points of η are included among those of η_* we have

$$F_\phi\{M(\eta_*), M(\eta)\} = \sum \eta_i F_\phi\{M(\eta_*), x_{(i)} x_{(i)}^T\}$$
$$= 0,$$

by Corollary 3.10. In this case

$$f(\eta) - f(\eta_*) = O(\|\eta_* - \eta\|^2).$$

However if any support point $x_{(i)}$ of η is such that $F_\phi\{M(\eta_*), x_{(i)} x_{(i)}^T\} < 0$, then $F_\phi\{M(\eta_*), M(\eta)\} < 0$ and

$$f(\eta) - f(\eta_*) = O(\|\eta_* - \eta\|).$$

If η is a design measure corresponding to an N-observation design, so that each η_i has the form r_i/N, this means that if this design includes only support points of an optimal design measure, then

$$\phi\{M(\eta_*)\} - \phi\{M(\eta)\} = O(N^{-2}),$$

whereas if it includes any point x for which $F_\phi\{M(\eta_*), xx^T\} < 0$, then

$$\phi\{M(\eta_*)\} - \phi\{M(\eta)\} = O(N^{-1}).$$

This suggests that, at least in the differentiable case, we would lose little by considering only N-observation designs concentrated on the support points of an optimal design measure. If we do so, and the number of support points of the optimal measure is finite then the N-observation design problem becomes much simpler since we have effectively reduced the design space to a finite set of points. Exchange algorithms now become a possibility, where we start with an arbitrary N-observation design with matrix M_0 and exchange one point of this design with another point in the finite design space to yield a new N-observation design with information matrix M_1; the points involved in the exchange may be chosen, for instance, to ensure that $\phi(M_1) - \phi(M_0)$ is maximal. We continue exchanging points in this way until no further increase in ϕ is possible.

Such exchange algorithms can be employed, of course, even when the design space is not finite, but their implementation then must depend on the possibility of obtaining an analytic, rather than an enumerative way of making the exchange so that $\phi(M_1) - \phi(M_0)$ is maximal; see Fedorov (1972, p. 164) for a discussion of this for D-optimality.

Finally we note that such algorithms do not necessarily yield an optimal N-observation design, since we may reach a stage where we cannot increase ϕ by exchanging one point, but might be able to do so by exchanging two or more at a time.

5
APPROXIMATE THEORY—PARTICULAR CRITERIA

As indicated earlier, approximate theory was developed for particular criteria before the unification provided by the general theory discussed in Chapter 3 was appreciated. Intensive study of particular criteria has, of course, resulted in additional particular results for them and in this chapter we shall discuss some of these.

5.1 D-optimality

5.1.1 The most intensively studied criterion is the D-optimal one for which

$$\phi(M) = \log \det M, \quad \text{if } M \text{ is non-singular}$$
$$= -\infty, \quad \text{otherwise};$$

see Section 2.2.1. Since there exist design measures η for which $M(\eta)$ is non-singular only when the design space $\mathscr{X}$ spans R^k, the D-optimal problem is non-degenerate only in this case.

One of the more appealing features of this criterion is that the extremum problem remains invariant under non-singular linear transformations of the design space. For if $\tilde{x}$ is a random vector with distribution η and $\tilde{z} = P\tilde{x}$, where P is a non-singular $k \times k$ matrix, then

$$M_{\tilde{z}}(\eta) = E(\tilde{z}\tilde{z}^T) = PE(\tilde{x}\tilde{x}^T)P^T = PM_{\tilde{x}}(\eta)P^T,$$

and

$$\log \det M_{\tilde{z}}(\eta) = \log (\det P)^2 + \log \det M_{\tilde{x}}(\eta).$$

Hence determining η to maximize $\log \det M_{\tilde{z}}(\eta)$ is the same problem as that of determining it to maximize $\log \det M_{\tilde{x}}(\eta)$.

We have already seen in Example 3.8 that, with ϕ defined as above,

$$F_\phi(M, xx^T) = x^T M^{-1} x - k$$

on $\mathscr{M}^+$ and that, because of this, D- and G-optimality are equivalent *for design measures*; see Section 3.11.

5.1.2 Silvey (1972) suggested the possibility of a duality between the D-optimal design measure problem and what has come to be called the minimum-ellipsoid problem and this duality was established by Sibson (1972), using Strong Lagrangian theory; see, for instance, Whittle (1971, pp. 41–67). The underlying geometry provides some additional insight into the D-optimal problem and we shall now discuss it briefly for this reason.

The minimal ellipsoid problem for a compact set $\mathscr{X}$ in R^k is: find an ellipsoid, centred on the origin, containing $\mathscr{X}$ and having minimal content. From geometrical considerations it is clear that such an ellipsoid will pass through certain boundary points of $\mathscr{X}$, since any ellipsoid containing $\mathscr{X}$ which does not do so can be shrunk to a new one which does and still contains $\mathscr{X}$. It is equally clear that unless the boundary of $\mathscr{X}$ has an ellipsoidal segment, then the minimal ellipsoid will pass through only a discrete, and usually finite, set of points of $\mathscr{X}$.

The duality theorem established by Sibson (1972) shows that the minimal ellipsoid for $\mathscr{X}$ is $\{z \in R^k : z^T M_*^{-1} z = k\}$, where M_* is the information matrix of a D-optimal design measure η_* on $\mathscr{X}$. Note that M_* is unique whether or not η_* is, since log det is strictly concave; see Section 3.3.2. Because of Corollary 3.10 the only possible support points of an optimal η_* are points of $\mathscr{X}$ through which the minimal ellipsoid passes since all other $x \in \mathscr{X}$ are such that $x^T M_*^{-1} x < k$, that is, such that $F_\phi(M_*, xx^T) < 0$.

The following very simple example illustrates these points. Suppose that $E(\tilde{y}|u, \theta) = \theta_1 u + \theta_2 u^2$ and that the real design space $\mathscr{U}$ is $[0, 1]$. Then $\mathscr{X}$ is the segment of the parabola $x_2 = x_1^2$ between $(0, 0)$ and $(1, 1)$. Geometry suggests that the minimum ellipse for this $\mathscr{X}$ will pass through $(1, 1)$ and touch the parabolic segment $\mathscr{X}$ at a point (a, a^2) where $0 < a < 1$. If this is so, then a D-optimal design measure has these two points only as support points; it must have both to have a non-singular information matrix. Given these suggestions from the geometry it is not difficult to show that the appropriate value of a is $\frac{1}{2}$ and that the design measure that puts probability $\frac{1}{2}$ at each of $(\frac{1}{2}, \frac{1}{4})$ and $(1, 1)$ is D-optimal. In this case, of course, the optimal design measure is unique.

A simple example which is interesting theoretically is the two-dimensional one where the design space $\mathscr{X}$ is itself an ellipse centred on the origin. Then *every* point in the design space is potentially an optimal design support point. We leave it to the reader to show that there is now an infinite number of optimal designs, including an infinite number all of which have two-point support. In this connection, and others, the following result is useful.

5.1.3 **Lemma.** *If $\mathscr{X} \in R^k$ and spans R^k, and if a D-optimal design measure is supported on k points, then it puts probability k^{-1} at each of them.*

Proof. We note that, in general, if η is a design measure with finite support $x_{(1)}, \ldots, x_{(m)}$ then

$$M(\eta) = \sum \eta_i x_{(i)} x_{(i)}^T = X^T D_\eta X$$

where X is the $m \times k$ matrix whose ith row is $x_{(i)}^T$ and D_η is diag $(\eta_1, \ldots, \eta_m)$. When $m = k$,

$$\det M(\eta) = (\det X)^2 \prod_{i=1}^{k} \eta_i,$$

and for non-singular X this is maximized, subject to $\eta_i \geq 0$ and $\sum \eta_i = 1$, by $\eta_i = k^{-1}$, $i = 1, \ldots, k$.

5.1.4 *Analytic results.* Earlier we have emphasized the importance of numerical methods for constructing optimal design measures. On occasion, however, it is possible to obtain them analytically and, in this context Lemma 5.1.3 is important. We illustrate by considering polynomial regression when we have a single control variable u and

$$E(\tilde{y}|u, \theta) = \theta_0 + \theta_1 u + \ldots + \theta_{k-1} u^{k-1}.$$

Suppose that the design space $\mathscr{U}$ is $[-1, 1]$; $\mathscr{X}$ is a subset of R^k with typical vector x given by $x^T = (1 \quad u \quad \ldots \quad u^{k-1})$, $-1 \leq u \leq 1$.

General theory tells us that:

(i) a design measure η_* is D-optimal iff $x^T \{M(\eta_*)\}^{-1} x \leq k$ for all $x \in \mathscr{X}$;

(ii) $x^T\{M(\eta_*)\}^{-1}x$ takes its maximum value of k at the support points of η_*.

Now when $x^T = (1 \quad u \quad \ldots \quad u^{k-1})$, $x^T\{M(\eta_*)\}^{-1}x$ is a polynomial in u of degree $2k-2$. A polynomial of degree $2k-2$ cannot have more than k maxima in the interval $[-1, 1]$; and if it has k, two of these must be at $u = -1$ and $u = +1$, since there cannot be more than $2k-3$ stationary values and there must be a local minimum between every two maxima. Hence a D-optimal design cannot have more than k support points. But it must have at least k to have a non-singular information matrix. It therefore has exactly k and two of these must be the x's corresponding to $u = -1$ and $u = +1$. By Lemma 5.1.3 it puts probability k^{-1} at each support point.

Let the support points of η_* be the x's corresponding to $u = \pm 1, u = u_1, \ldots, u = u_{k-2}$. Then $M(\eta_*) = (1/k)X^TX$, where

$$X = \begin{pmatrix} 1 & 1 & \ldots & 1 \\ 1 & -1 & \ldots & (-1)^{k-1} \\ 1 & u_1 & \ldots & u_1^{k-1} \\ \vdots & & & \\ 1 & u_{k-2} & \ldots & u_{k-2}^{k-1} \end{pmatrix},$$

and

$$\det M(\eta_*) = \frac{1}{k^k}(\det X)^2 \propto \left\{\prod_{i=1}^{k-2}(1-u_i^2)\prod_{i<j}(u_i-u_j)\right\}^2.$$

Now the D-optimal problem reduces to that of determining $u_1, \ldots, u_{k-2}$ to maximize the right-hand expression. It can be shown by elementary calculus that the u_i's are the roots of $P'_{k-1}(u)$, where $P_k(u)$ is the kth Legendre polynomial.

For details of this and similar results, see Fedorov (1972, p. 83).

There is a vast literature on D-optimality and here we have focused attention on only a few aspects of it.

5.2 D_A- and D_s-optimality

5.2.1 We recall the definition of the D_A-optimal criterion function, already discussed in Section 3.13. Let A^T be an $s \times k$ matrix of rank $s < k$, and let $\mathcal{M}_A$ be the subset of $\mathcal{M}$ consisting of those matrices M such that $A = MY$ for some Y. Define

$$\phi_A(M) = -\log \det A^T M^- A, \quad \text{if } M \in \mathcal{M}_A$$
$$= -\infty, \quad \text{otherwise}.$$

Here M^- is any g-inverse of M; of course if M is non-singular, $M^- = M^{-1}$.

The calculation of $G_{\phi_A}(M_1, M_2)$ is relatively straightforward when M_1 is non-singular and we find that then

$$G_{\phi_A}(M_1, M_2) = \text{tr } \{M_1^{-1} A (A^T M_1^{-1} A)^{-1} A^T M_1^{-1} M_2\}$$

and that

$$F_{\phi_A}(M_1, M_2) = G_{\phi_A}(M_1, M_2 - M_1) = G_{\phi_A}(M_1, M_2) - s.$$

Since, at non-singular M_1, G_{ϕ_A} is linear in its second argument, Theorem 3.7 is applicable and we have the result that a design measure η_* with *non-singular* information matrix M_* is D_A-optimal if and only if

$$F_{\phi_A}(M_*, xx^T) \leqslant 0, \quad \text{for all } x \in \mathcal{X},$$

that is, if and only if,

$$x^T M_*^{-1} A (A^T M_*^{-1} A)^{-1} A^T M_*^{-1} x \leqslant s, \quad \text{for all } x \in \mathcal{X}.$$

Suppose however that η_* has *singular* information matrix M_r of rank $r < k$. Then as in Section 3.13, η_* is D_A-optimal if and only if there exists a matrix $H \in \mathcal{H}(M_r)$ and such that

$$x^T (M_r + HH^T)^{-1} A \{A^T (M_r + HH^T)^{-1} A\}^{-1}$$
$$\times A^T (M_r + HH^T)^{-1} x \leqslant s, \quad \text{for all } x \in \mathcal{X}.$$

This condition is, of course, much more difficult to verify than that for non-singular M_*.

Calculation of $F_{\phi_A}(M_r, M)$ for singular $M_r \in \mathcal{M}_A$ is tedious for general M, but it is probably worth recording that

$$F_{\phi_A}(M_r, xx^T) = -s \quad \text{for any } x \text{ not in the range of } M_r;$$

while, if H is any matrix in $\mathcal{H}(M_r)$, and x *is* in the range of M_r

$$F_{\phi_A}(M_r, xx^T) = F_{\phi_A}(M_r + HH^T, xx^T);$$

see Silvey (1978).

5.2.2

Much attention has been paid in the literature to the case where $A^T = (I_s \ 0)$, that is to the case of D_s-optimality. We shall denote the corresponding criterion function by ϕ_s. Of course the above results for general A apply to this case also; however it is possible to obtain alternative formulations of them.

We partition M into

$$\begin{pmatrix} M_{11} & M_{12} \\ M_{12}^T & M_{22} \end{pmatrix},$$

where M_{11} is $s \times s$.

Let C be any solution of the matrix equation $CM_{22} = M_{12}$; this equation necessarily has a solution; see Karlin & Studden (1966). Then $M \in \mathcal{M}_A$ if and only if $M_{11} - CM_{22}C^T$ is positive definite. This latter matrix does not depend on which solution C is chosen and the definition of ϕ_s may be formulated alternatively as:

$$\phi_s(M) = \log \det (M_{11} - CM_{22}C^T), \quad \text{when } M_{11} - CM_{22}C^T \text{ is non-singular}$$
$$= -\infty, \quad \text{otherwise.}$$

If M is non-singular then so is M_{22}; C is then unique and

$$\phi_s(M) = \log \det (M_{11} - M_{12}M_{22}^{-1}M_{12}^T).$$

In this case

$$G_{\phi_s}(M, xx^T) = (x^{(1)} - M_{12}M_{22}^{-1}x^{(2)})^T (M_{11} - M_{12}M_{22}^{-1}M_{12}^T)^{-1}$$
$$\times (x^{(1)} - M_{12}M_{22}^{-1}x^{(2)}),$$

where x has been partitioned, in the obvious way,

$$x = \begin{pmatrix} x^{(1)} \\ x^{(2)} \end{pmatrix}.$$

A design measure with *non-singular* information matrix M is D_s-optimal if and only if

$$(x^{(1)} - M_{12}M_{22}^{-1}x^{(2)})^T (M_{11} - M_{12}M_{22}^{-1}M_{12}^T)^{-1}$$
$$\times (x^{(1)} - M_{12}M_{22}^{-1}x^{(2)}) \leqslant s \quad \text{for all } x \in \mathcal{X}.$$

An alternative formulation of our general condition above that a design measure with *singular* information matrix M be D_s-optimal is that there exists a matrix C satisfying $CM_{22} = M_{12}$

and such that

$$(x^{(1)} - Cx^{(2)})^T(M_{11} - CM_{22}C^T)^{-1}(x^{(1)} - Cx^{(2)}) \leq s$$
for all $x \in \mathscr{X}$.

Note that while it takes some algebra to demonstrate the fact, this is simply an alternative formulation of the D_A-optimality condition for the case where $A^T = (I_s \quad 0)$.

This result was first established by Karlin & Studden (1966), though there was a slight fallacy in their argument, subsequently corrected by Atwood (1969). An alternative proof using Strong Lagrangian theory, rather than game theory, was given by Silvey & Titterington (1973). The geometry lying behind this alternative approach is quite illuminating and we shall now discuss it briefly.

5.2.3 *The thinnest s-cylinder problem.*

For D-optimality we posed a primal problem, the minimum-ellipsoid problem, of which the D-optimal design measure problem was the dual. We can do the same for D_s-optimality. Different primal problems may be chosen; see Sibson (1974), Pukelsheim (1980). We choose here what we term the thinnest s-cylinder problem, defined as follows: given a compact set $\mathscr{X}$ in R^k, find an $s \times (k-s)$ matrix B and a positive definite $s \times s$ matrix A to maximize $\det A$ subject to

$$(x^{(1)} - Bx^{(2)})^T A(x^{(1)} - Bx^{(2)}) \leq s \quad \text{for all } x \in \mathscr{X}.$$

The reason for calling this a thinnest cylinder problem can best be seen by looking at the case $k = 3$, $s = 2$. Let $A = (a_{ij})$ be a positive definite 2×2 matrix, and b a real number. Then the set

$$\{z \in R^3 : a_{11}(z_1 - bz_3)^2 + 2a_{12}(z_1 - bz_3)(z_2 - bz_3) + a_{22}(z_2 - bz_3)^2 = 2\}$$

is an elliptic cylinder in R^3 whose axis, namely the line $z_1 = z_2 = bz_3$, passes through the origin. Its section by the plane $z_3 = 0$ is an ellipse with area proportional to $(\det A)^{-1/2}$. Given a compact set $\mathscr{X}$ in R^3 we are seeking to find a cylinder of this form, containing $\mathscr{X}$ and having minimal cross-section area. The generalization of this geometry to higher dimensions is obvious.

Silvey & Titterington (1973) have proved that the D_s-optimal design measure problem for $\mathscr{X}$ is the dual of the thinnest s-cylinder problem for $\mathscr{X}$. This does not help in the construction of D_s-optimal designs, but the geometry involved once again provides some additional insight into the problem and aids in the understanding of the results quoted in Section 5.2.2. For instance it is clear that the thinnest s-cylinder for $\mathscr{X}$ will pass through some points of $\mathscr{X}$ and strictly contain others. It emerges from the duality theorem that only those points of $\mathscr{X}$ through which the thinnest cylinder passes are potential support points of a D_s-optimal design measure. The thinnest cylinder problem involves determining both the 'shape' of the cylinder (the matrix A) and the 'direction' of its axis (the matrix B). This 'explains' the matrix C in Section 5.2.2.

A very simple example, considered earlier, illustrates these points. Let $E(\tilde{y}|x, \theta) = \theta_1 x_1 + \theta_2 x_2$ and let $\mathscr{X}$ be the quadrilateral with vertices $(0, 0), (1, 0), (4, 1), (4, 2)$. We ask the question: is the design measure η_1 that puts probability 1 at $(1, 0)$, D_1-optimal, that is, optimal for estimating θ_1. The duality established in the Silvey-Titterington theorem enables us to rephrase this question: can we, from the information matrix

$$M_1 = \begin{pmatrix} 1 & 0 \\ 0 & 0 \end{pmatrix}$$

of η_1, generate a cylinder—a pair of parallel lines in this instance —that contains $\mathscr{X}$? In other words can we find a c satisfying $c \times 0 = 0$ such that the cylinder $(z_1 - cz_2)^2 = 1$ contains $\mathscr{X}$? Note that in this example $M_{12} = M_{22} = 0$ and $M_{11} - cM_{22}c^T = 1$. Clearly we cannot; the best hope is $c = 3$, corresponding to the line joining $(1, 0)$ and $(4, 1)$; however the line $z_1 - 3z_2 = -1$ meets $z_1 = 4$ where $z_2 = \frac{5}{3}$, and the cylinder $(z_1 - 3z_2)^2 = 1$ does not contain $\mathscr{X}$. Therefore η_1 is *not* D_1-optimal.

The geometry also makes it clear that the thinnest cylinder for $\mathscr{X}$ can pass through none of its points other than $(4, 1)$ and $(4, 2)$ which are therefore the only potential support points of a D_1-optimal measure; and they must both be support points since no design supported on only one of them admits estimation of θ_1.

For this particular example the alternative formulation of the necessary and sufficient condition, stated in Section 5.2.1, is as follows. Matrices $H \in \mathscr{H}(M_1)$ are vectors

$$h = \begin{pmatrix} h_1 \\ h_2 \end{pmatrix} \quad \text{with } h_2 \neq 0.$$

Does there exist such an h with

$$x^T(M_1 + hh^T)^{-1}\begin{pmatrix}1\\0\end{pmatrix}\left\{(1\ \ 0)(M_1 + hh^T)^{-1}\begin{pmatrix}1\\0\end{pmatrix}\right\}^{-1}$$
$$\times (1\ \ 0)(M_1 + hh^T)^{-1}x \leq 1$$

for all $x \in \mathscr{X}$? If we substitute for M_1 and simplify, this becomes

$$\left(x_1 - \frac{h_1}{h_2}x_2\right)^2 \leq 1, \quad \text{for all } x \in \mathscr{X}.$$

This demonstrates the equivalence of the two formulations, an equivalence which, of course, obtains generally. In this case the vector h determines the direction of the axis of the cylinder and this is the role played in general for D_s-optimality by the matrix H in $\mathscr{H}(M_r)$.

5.3 *A linear criterion function*

Section 2.2.5 motivated consideration of the function ϕ defined, for singular M by

$$\phi(M) = -\text{tr}\,(A^T M^{-1} A),$$

where A is a $k \times s$ matrix of rank s; and in Section 3.13 we extended the definition of ϕ to the class $\mathscr{M}_A$, that is, matrices M such that $A = MY$, for some Y. We now consider this criterion function, which we shall denote by ϕ_L, defined by

$$\phi_L(M) = -\text{tr}\,(A^T M^- A), \quad \text{if } M \in \mathscr{M}_A$$
$$= -\infty, \quad \text{otherwise}.$$

This function also is differentiable at non-singular M, but not differentiable at singular M. If M_1 is non-singular, a simple calculation shows that

$$G_{\phi_L}(M_1, M_2) = \text{tr}\,(A^T M_1^{-1} M_2 M_1^{-1} A),$$

and that

$$F_{\phi_L}(M_1, M_2) = G_{\phi_L}(M_1, M_2) - \text{tr}\,(A^T M_1^{-1} A).$$

In particular

$$F_{\phi_L}(M_1, xx^T) = x^T M_1^{-1} A A^T M_1^{-1} x - \text{tr}\,(A^T M_1^{-1} A)$$

APPROXIMATE THEORY—PARTICULAR CRITERIA

and a necessary and sufficient condition that a design measure η_* with non-singular information matrix M_* be ϕ_L-optimal is that

$$x^T M_*^{-1} A A^T M_*^{-1} x \leq \mathrm{tr}\,(A^T M_*^{-1} A) \qquad \text{for all } x \in \mathcal{X}.$$

As indicated in Section 3.13 if a design measure η_* has *singular* information matrix M_r of rank $r < k$ and $M_r \in \mathcal{M}_A$, then η_* is ϕ_L-optimal if and only if there exists an $H \in \mathcal{H}(M_r)$ and such that

$$x^T (M_r + HH^T)^{-1} A A^T (M_r + HH^T)^{-1} x$$
$$\leq \mathrm{tr}\,\{A^T (M_r + HH^T)^{-1} A\}, \qquad \text{for all } x \in \mathcal{X}.$$

Once again there seems to be little point in giving an explicit expression for $F_{\phi_L}(M_r, M)$ for general M. However we note that if x is not in the range of M_r then

$$F_{\phi_L}(M_r, xx^T) = \phi_L(M_r),$$

while if x is in the range of M_r,

$$F_{\phi_L}(M_r + HH^T, xx^T) = F_{\phi_L}(M_r, xx^T), \qquad \text{for all } H \in \mathcal{H}(M_r).$$

5.4 c-optimality

The criterion of c-optimality, originally considered by Elfving (1952), is the special case of the above linear criterion when the matrix A is a column vector c. Consequently the above formula apply with appropriate simplification. For example a design measure η_r with singular information matrix M_r is c-optimal if and only if there exists an $H \in \mathcal{H}(M_r)$ such that

$$\{x^T (M_r + HH^T)^{-1} c\}^2 \leq c^T (M_r + HH^T)^{-1} c \qquad \text{for all } x \in \mathcal{X}.$$

We note also that c-optimality can be considered as a particular case of D_A-optimality.

For some suggestive geometry involving the convex hull of $\mathcal{X} \cup (-\mathcal{X})$, see Elfving (1952) and Pukelsheim (1979).

5.5 *Examples*

5.5.1 Suppose that $k = 2$ and the design space $\mathcal{X}$ is the triangle with vertices $(0, 0)$, $(1, 0)$ and $(0, 1)$; we wish to design in order to predict $E(\tilde{y})$ over the triangle $\triangle$ with vertices $(1, 0)$, $(2, 0)$ and $(0, 1)$; and we wish the average variance of predicted values to be as small as possible, this average being taken with respect to the uniform distribution on the triangle.

We note first that

$$\iint_\Delta cc^T dc_1 dc_2 = \frac{1}{24}\begin{pmatrix} 14 & 3 \\ 3 & 2 \end{pmatrix}.$$

Hence the average value of cc^T with respect to the uniform distribution on the triangle is

$$B = \frac{1}{12}\begin{pmatrix} 14 & 3 \\ 3 & 2 \end{pmatrix},$$

and our criterion function is

$$\phi(M) = -\operatorname{tr}(M^{-1}B);$$

see Section 2.2.5. Since B is non-singular, only designs with non-singular information matrices are of interest. This is intuitively obvious from a practical point of view since only such designs will enable us to predict $E(\tilde{y})$ over all points of the triangle. From a mathematical point of view, when we express B in the form AA^T, A is non-singular, and if $A = MY$ for some Y, then M is necessarily non-singular. In other words the class $\mathcal{M}_A$ is the class of non-singular information matrices.

Now we know from Corollary 3.10 that if M_* is the information matrix of an optimal measure η_*, then $F_\phi(M_*, xx^T)$ takes its maximum value over $\mathcal{X}$ at the support points of η_*. But

$$F_\phi(M_*, xx^T) = x^T M_*^{-1} B M_*^{-1} x - \operatorname{tr}(M_*^{-1}B)$$

and this is a strictly convex function of x. Hence it takes its maximum value only at extreme points of the convex set $\mathcal{X}$. It is obvious that $(0, 0)$ cannot be a support point of η_* and therefore its only possible support points are $(1, 0)$ and $(0, 1)$. Both of these must be support points since M_* must be non-singular. Therefore M_* has the form

$$M_* = \begin{pmatrix} p & 0 \\ 0 & 1-p \end{pmatrix},$$

and $\operatorname{tr}(M_*^{-1}B)$ is proportional to $7/p + 1/(1-p)$, which is minimal at $p = \frac{1}{6}(7 - \sqrt{7})$. The optimal design measure therefore puts this probability at $(1, 0)$ and probability $\frac{1}{6}(\sqrt{7} - 1)$ at $(0, 1)$. It can now be verified that for this design measure we do indeed have $F_\phi(M_*, xx^T) \leqslant 0$ for all $x \in \mathscr{X}$, with equality only when $x^T = (1, 0)$ and $(0, 1)$.

5.5.2   In contrast suppose that with the same design space as in Section 5.5.1 we wish to estimate $\theta_1 + \theta_2$ with minimum variance; that is we are interested in $c^T\theta$ with $c^T = (1 \;\; 1)$. The geometric argument used by Elfving (1952) shows that the design measure which puts probability $\frac{1}{2}$ at each of $(1, 0)$ and $(0, 1)$ is optimal. However practical intuition suggests that the measure η_* which puts probability 1 at $(\frac{1}{2}, \frac{1}{2})$ is also optimal. This design measure has the *singular* information matrix

$$M_* = \frac{1}{4}\begin{pmatrix} 1 & 1 \\ 1 & 1 \end{pmatrix}$$

and we consider it for this reason. We recall that η_* is optimal if and only if there exists $H \in \mathscr{H}(M_*)$ and such that

$$\{x^T(M_* + HH^T)^{-1}c\}^2 \leqslant c^T(M_* + HH^T)^{-1}c, \quad \text{for all } x.$$

In this case, matrices $H \in \mathscr{H}(M_*)$ are vectors

$$h = \begin{pmatrix} h_1 \\ h_2 \end{pmatrix}, \quad \text{with } h_2 \neq 0.$$

In attempting to verify whether there does exist an h such that the condition is satisfied, the natural h to try first is one in the null-space of M_*, that is, one which is orthogonal to the range of M_*. So we try

$$h = \frac{1}{4}\begin{pmatrix} 1 \\ -1 \end{pmatrix}.$$

Here the $\frac{1}{4}$ is introduced for arithmetic convenience; as pointed out by Silvey (1978) every vector of the form (non-zero constant) $\times \begin{pmatrix} 1 \\ -1 \end{pmatrix}$ is equivalent as far as the condition we are setting out to verify is concerned.

With h as above and $c^T = (1 \;\; 1)$, we find that

$$x^T(M_* + hh^T)^{-1}c = 2(x_1 + x_2),$$

and

$$c^T(M_* + hh^T)^{-1}c = 4.$$

OPTIMAL DESIGN

For all $x \in \mathcal{X}$, $\{2(x_1 + x_2)\}^2 \leq 4$ and the design measure putting probability 1 at $(\frac{1}{2}, \frac{1}{2})$ is indeed optimal.

We note in passing that $\{2(x_1 + x_2)\}^2 = 4$ at all points on the line segment joining $(1, 0)$ and $(0, 1)$. This suggests that all such points are potential support points of an optimal design measure for estimating $\theta_1 + \theta_2$. The reader may verify that there is in fact an infinity of optimal design measures for this problem, including the continuous design measure which is uniform on the line segment joining $(1, 0)$ and $(0, 1)$.

5.6 *Other criteria*

The particular criteria considered in this chapter are those which feature most prominently in the literature, and which are of most practical importance. Kiefer (1974) considers a general class of criterion functions, a class which includes all those considered here. The reader may consult his paper for further details, and also a paper by Pukelsheim (1980) who establishes duality theorems for this class.

6
NON-LINEAR PROBLEMS

6.1 Introduction

We return now to the general situation discussed in Chapter 1, but dropping the distinction between parameters of interest and nuisance parameters; discussion of linear theory will have made it clear that this distinction only enters the picture in the definition of the criterion function and from now on we shall take account of it automatically in this definition.

Specifically, then, the probability density function $p(y|u, \theta)$ of an observable random variable $\tilde{y}$ depends on a vector u of control variables and a k-vector θ of unknown parameters; we are given a compact design space $\mathcal{U}$ from which we may choose vectors u at which to observe $\tilde{y}$, that is, we can choose different designs; and the object of design is to estimate θ, or certain functions of θ, 'as well as possible'.

We recall that for a single observation on $\tilde{y}$ at u, Fisher's information matrix is $J(u, \theta)$, the $k \times k$ matrix whose (i,j)th element is $E\{-\partial^2 \log p(\tilde{y}|u, \theta)/\partial\theta_i\partial\theta_j\}$ and that for n independent observations at $u_{(1)}, \ldots, u_{(n)}$, the corresponding matrix is $L(\mathbf{u}, \theta) = \sum_i J(u_{(i)}, \theta)$, where $\mathbf{u} = (u_{(1)}, \ldots, u_{(n)})$.

It is convenient to retain the notion of a design measure, but now regarded as a probability distribution μ on the actual design space $\mathcal{U}$ as opposed to the induced design space $\mathcal{X}$. The information matrix $M(\mu, \theta)$ of μ is defined as $E\{J(\tilde{u}, \theta)\}$ where $\tilde{u}$ is a random vector with distribution μ. Thus if μ is the design measure associated with the n-observation design $\mathbf{u} = (u_{(1)}, \ldots, u_{(n)})$, that is, the distribution that puts probability $1/n$ at each of $u_{(1)}, \ldots, u_{(n)}$, then

$$M(\mathbf{u}, \theta) = \frac{1}{n} L(\mathbf{u}, \theta).$$

Approximate theory is concerned with finding $\mu_*(\theta)$ to maximize some function $\phi\{M(\mu, \theta)\}$ or, more generally, a function $\phi\{M(\mu, \theta), \theta\}$, that is, a function which may depend on θ

other than through $M(\mu, \theta)$ only. We shall say that such a $\mu_*(\theta)$ is ϕ_θ-optimal.

The essential simplification that occurs in the linear case is that the maximizing $\mu_*(\theta)$ does not depend on θ. Investigation of the way in which $\mu_*(\theta)$ does depend on θ in the non-linear case will guide our thinking about what should be done in practice.

Immediately, for each fixed value of θ, we have analogues of the important Theorems 3.6 and 3.7. For given θ, let $\mathcal{M}_\theta$ denote the set of information matrices generated as μ ranges over the class of all probability distributions on $\mathcal{U}$, including those which assign probability 1 to single points of $\mathcal{U}$. Then $\mathcal{M}_\theta$ is the convex hull of $\{J(u, \theta); u \in \mathcal{U}\}$. The analogue of Theorem 3.6 is:

6.1.1 Theorem. *If θ is fixed and ϕ is concave on $\mathcal{M}_\theta$, $\mu_*(\theta)$ is ϕ_θ-optimal if and only if $F_\phi[M\{\mu_*(\theta), \theta\}, M(\mu, \theta)] \leqslant 0$ for all μ.*

Similarly we have:

6.1.2 Theorem. *If θ is fixed and ϕ is concave on $\mathcal{M}_\theta$ and differentiable at $M\{\mu_*(\theta), \theta\}$, then $\mu_*(\theta)$ is ϕ_θ-optimal if and only if $F_\phi[M\{\mu_*(\theta), \theta\}, J(u, \theta)] \leqslant 0$ for all $u \in \mathcal{U}$.*

The proofs of these theorems exactly parallel those of Theorems 3.6 and 3.7 respectively and need not be repeated.

Equally it is true that algorithms analogous to those of Chapter 4 can be used to construct ϕ_θ-optimal design measures for each given θ. We re-emphasize, however, that ϕ_θ-optimal design measures corresponding to different θ will typically be different and it is this fact that raises new *practical* problems.

We turn now to some examples to illustrate possibilities.

6.2 Example

This example is due to White (1975).
An experimenter may observe:

(i) a Poisson random variable with mean θ_1;
(ii) a Poisson random variable with mean θ_2;
(iii) a Poisson random variable with mean $\theta_1 + \theta_2$.

To conform with our general notation we introduce a control

variable u which takes the value u_1 if he chooses to observe (i), u_2 if (ii) and u_3 if (iii). Then, for $y = 0, 1, 2, \ldots$,

$$p(y|u_1, \theta) = e^{-\theta_1}\theta_1^y/y!$$
$$p(y|u_2, \theta) = e^{-\theta_2}\theta_2^y/y!$$
$$p(y|u_3, \theta) = e^{-(\theta_1+\theta_2)}(\theta_1 + \theta_2)^y/y!,$$

where $\theta^T = (\theta_1 - \theta_2)$. A trivial calculation shows that

$$J(u_1, \theta) = \begin{pmatrix} \theta_1^{-1} & 0 \\ 0 & 0 \end{pmatrix}$$

$$J(u_2, \theta) = \begin{pmatrix} 0 & 0 \\ 0 & \theta_2^{-1} \end{pmatrix}$$

$$J(u_3, \theta) = (\theta_1 + \theta_2)^{-1}\begin{pmatrix} 1 & 1 \\ 1 & 1 \end{pmatrix}.$$

For the design measure μ which puts probability μ_i at u_i, $i = 1, 2, 3$,

$$M(\mu, \theta) = \begin{pmatrix} \mu_1\theta_1^{-1} + \mu_3(\theta_1+\theta_2)^{-1} & \mu_3(\theta_1+\theta_2)^{-1} \\ \mu_3(\theta_1+\theta_2)^{-1} & \mu_2\theta_2^{-1} + \mu_3(\theta_1+\theta_2)^{-1} \end{pmatrix}.$$

Now suppose that $\phi = \log \det$ and consider the design measure μ_* that puts probability $\frac{1}{2}$ at each of u_1 and u_2, so that

$$M(\mu_*, \theta) = \frac{1}{2}\begin{pmatrix} \theta_1^{-1} & 0 \\ 0 & \theta_2^{-1} \end{pmatrix}.$$

Since $M(\mu_*, \theta)$ is non-singular, ϕ is differentiable there. Also

$$F_\phi\{M(\mu_*, \theta), J(u, \theta)\} = \text{tr}\{M^{-1}(\mu_*, \theta)J(u, \theta)\} - 2$$

and we have

$$F_\phi\{M(\mu_*, \theta), J(u_i, \theta)\} = 0, \quad i = 1, 2, 3.$$

Appeal to Theorem 6.1.2 now shows that for every θ, μ_* is D_θ-optimal.

This example is, of course, unusual in that while the information matrix $M(\mu, \theta)$ does depend on θ, the D_θ-optimal μ_* does not. The practical implications of the result we have established would scarcely surprise the practising statistician: if the experimenter is interested in estimating both θ_1 and θ_2, and he is equally interested in them, he can do no better than take half his observations from the Poisson distribution with mean θ_1 and half from that with mean θ_2!

6.3 Example

With the same set-up as in the previous example, suppose that the experimenter wishes to estimate $\theta_1 + \theta_2$ with minimum variance. It is intuitively obvious that his best policy is to take all observations from the Poisson distribution with mean $\theta_1 + \theta_2$; and this can be proved quite easily in a variety of ways. While it is cumbersome, it is also quite instructive to use the apparatus that we have developed to demonstrate that the design measure μ_* which puts probability 1 at u_3 is optimal for all θ_1 and θ_2.

We note that for fixed θ each $J(u, \theta)$ can be expressed in the form xx^T; for u_1, the appropriate x^T is $x_{(1)}^T = (\theta_1^{-1/2} \quad 0)$; for u_2 it is $x_{(2)}^T = (0 \quad \theta_2^{-1/2})$; and for u_3,

$$x_{(3)}^T = \{(\theta_1 + \theta_2)^{-1/2} \quad (\theta_1 + \theta_2)^{-1/2}\}.$$

Hence, for fixed θ, we can view the problem in the following way: an induced design space $\mathscr{X}_\theta$ consists of the three points $x_{(1)}$, $x_{(2)}$, $x_{(3)}$ in R^2; a design measure η is a probability distribution on these three points and has information matrix $M(\eta, \theta) = \sum_{j=1}^{3} \eta_j x_{(j)} x_{(j)}^T$. Writing $A^T = (1 \quad 1)$, our criterion function ϕ is given by

$$\phi(M) = -A^T M^- A, \quad \text{if } M \in \mathscr{M}_A$$
$$= -\infty, \quad \text{otherwise};$$

see Section 3.13. In this case there is only one singular M in $\mathscr{M}_A$, namely the information matrix M_1 of the design measure η_1 that puts probability 1 at $x_{(3)}$. Is η_1 ϕ_θ-optimal? By Theorem 3.13.1 it is if we can find an $H \in \mathscr{H}(M_1)$ such that $F_\phi(M_1 + HH^T, x_{(i)} x_{(i)}^T) \leq 0$ for $i = 1, 2, 3$. The geometrical considerations of Section 5.2.3 suggest that an appropriate H might be

$$H = (\theta_1 + \theta_2)^{-1/2} \begin{pmatrix} \theta_1^{-1/2} \\ -\theta_2^{-1/2} \end{pmatrix}.$$

Here $(\theta_1 + \theta_2)^{-1/2}$ is introduced purely for algebraic convenience;

NON-LINEAR PROBLEMS

see Section 5.5.2. For this H it is easily verified that

$$F_\phi(M_1 + HH^T, x_{(i)}x_{(i)}^T) = -\frac{2(\theta_1\theta_2)^{1/2}}{\theta_1 + \theta_2 + 2(\theta_1\theta_2)^{1/2}}, \quad i = 1, 2,$$

$$= 0, \quad i = 3.$$

Therefore η_1, is indeed optimal, and this is so for all θ.

This example demonstrates how sometimes we can use linear theory in the non-linear case by interpreting the problem as a linear one where the induced design space varies with the unknown parameter. We can do so when the $J(u, \theta)$ have rank 1 and we are interested in linear functions of θ.

6.4 Example

We turn now to an example where the model is linear but interest is centred on a non-linear parametric function.

6.4.1 Suppose that

$$E(\tilde{y}|u, \theta) = \theta_1 u + \theta_2 u^2,$$

where it is known that $\theta_1 \geq 0$ and $\theta_2 < 0$; as usual we assume that observation errors are uncorrelated and have common variance. Given the design space $\mathcal{U} = [-1, 1]$, we wish to design in order to estimate with minimum variance, the value of u where the regression function is maximal; that is we are interested in the non-linear function $g(\theta) = -\theta_1/2\theta_2$. We have

$$\begin{pmatrix} \dfrac{\partial g}{\partial \theta_1} \\ \dfrac{\partial g}{\partial \theta_2} \end{pmatrix} = \begin{pmatrix} -\dfrac{1}{2\theta_2} \\ \dfrac{\theta_1}{2\theta_2^2} \end{pmatrix} = -\frac{1}{2\theta_2}\begin{pmatrix} 1 \\ 2g \end{pmatrix} = -\frac{1}{2\theta_2} c_g,$$

where $c_g^T = (1 \quad 2g)$, and we write g instead of $g(\theta)$ for typographical convenience.

Let μ be a design measure on $\mathcal{U}$. Its information matrix is

$$M(\mu) = E\begin{pmatrix} \tilde{u}^2 & \tilde{u}^3 \\ \tilde{u}^3 & \tilde{u}^4 \end{pmatrix},$$

where $\tilde{u}$ is a random variable with distribution μ. As in Example 1.5, but now allowing for the possibility that a singular design is optimal for fixed θ the criterion function to be maximized is

$$-\frac{1}{4\theta_2^2} c_g^T M^-(\mu) c_g,$$

defined and finite for $M(\mu) \in \mathcal{M}_{c_g}$.

Equivalently we may take as criterion function

$$\phi\{M(\mu), g\} = -c_g^T M^-(\mu) c_g.$$

In this example the induced design space $\mathcal{X}$ in R^2 is the same for all θ and is a segment of the parabola $x_2 = x_1^2$. The geometric argument of Elfving (1952) may be used to show that, for fixed g, the optimal design measure $\mu_*(g)$ has support included in $u = \pm 1$ and that

$$P\{u = 1 | \mu_*(g)\} = \frac{1}{2} + \frac{1}{4g}, \qquad \text{if } g \geqslant \frac{1}{2}$$

$$= \frac{1}{2} + g, \qquad \text{if } 0 \leqslant g \leqslant \frac{1}{2}.$$

Thus the optimal design measure for estimating g *does* vary with g in this case. Of particular interest is the fact that $\mu_*(\frac{1}{2})$ has the *singular* information matrix all of whose components are 1 since $\mu_*(\frac{1}{2})$ puts probability 1 at $u = 1$.

6.4.2 How does this analysis guide our thinking about what we might do in practice if faced with this problem?

One possible course of action is to choose a design whose corresponding design measure is optimal for some particular g in the hope that it is not too bad whatever the true g happens to be. An obvious candidate is the design according to which we take half our observations at $u = 1$ and half at $u = -1$; the corresponding design measure is optimal for $g = 0$ and nearly optimal for large g.

We can gauge how relatively well this design measure, $\mu_*(0)$, performs for other values of g by calculating its *efficiency*, measured by the ratio $c_g^T M^-\{\mu_*(g)\} c_g / [c_g^T M^-\{\mu_*(0)\} c_g]$. A simple calculation shows that this efficiency is $(1 + 4g^2)^{-1}$ for $0 \leqslant g \leqslant \frac{1}{2}$ and $1 - (1 + 4g^2)^{-1}$ for $g \geqslant \frac{1}{2}$. Thus while $\mu_*(0)$ is fairly efficient over a wide range of values of g, its efficiency falls to as low as 50% at $g = \frac{1}{2}$.

NON-LINEAR PROBLEMS

6.4.3 Pursuing this line of thought we might think in terms of finding a *maximin* design measure, one whose minimum efficiency is greater than that of any other. A fairly simple calculation shows that the design measure putting probability $\frac{3}{4}$ at $u = 1$ and $\frac{1}{4}$ at $u = -1$ is maximin. Its minimum efficiency is 75%, which occurs at $g = 0$ and $g = \frac{1}{2}$; it is optimal for $g = \frac{1}{4}$ and $g = 1$.

6.4.4 In this example we have already assumed that we have some prior knowledge of θ, that in fact $\theta_1 \geqslant 0$ and $\theta_2 < 0$, which imply prior knowledge that $g \geqslant 0$. We may have more precise prior knowledge than this and if we have, the design which we use should make use of this prior knowledge in either an informal or a formal way.

Suppose, for instance, that we know *a priori* that g is close to $\frac{1}{2}$. Informal use of this knowledge would result in choosing a design in which the majority of observations were taken at $u = +1$, because in the vicinity of $g = \frac{1}{2}$, the optimal design measure puts probability near 1 at $u = +1$.

Formal use of prior knowledge would involve introducing a prior distribution for θ and adopting a Bayesian approach to the whole problem. As an alternative to introducing a utility function and treating the problem as a decision-theoretic one, we might try to find a design giving maximum expected decrease in the entropy of the distribution of g; see Lindley (1956). However we shall not pursue this line of attack since it is outside our main theme.

We now look at these various proposals for dealing with the practical problem in the light of another example.

6.6 Example. Linear logistic quantal response model

This example has been discussed by Ford (1976). In it the response y takes the values 0 and 1; the control variable u is real; $\theta^T = (\theta_1 \quad \theta_2)$ and

$$p(\tilde{y} = 1 | u, \theta) = 1 - p(\tilde{y} = 0 | u, \theta)$$
$$= \exp(\theta_1 + \theta_2 u) / \{1 + \exp(\theta_1 + \theta_2 u)\}.$$

Let us suppose that the design space $\mathcal{U}$ is $[-1, 1]$ and that it is known that $\theta_1 > 0$ and $\theta_2 > 0$.

Ford has shown that the D_θ-optimum design measure is supported on two points at each of which it puts probability $\frac{1}{2}$. The support points depend on θ in the following way.

Let a be the positive solution of the equation $e^z = (z+1)/(z-1)$; $a \doteq 1.5434$. Also let

$$\Theta_1 = \{\theta : \theta_1 > 0, \theta_2 > 0, \theta_2 - \theta_1 \geq a\}$$
$$\Theta_2 = \{\theta : \theta_1 > 0, \theta_2 > 0, \theta_2 - \theta_1 < a, \exp(\theta_1 + \theta_2) \leq (\theta_2 + 1)/(\theta_2 - 1)\}$$
$$\Theta_3 = \{\theta : \theta_1 > 0, \theta_2 > 0, \exp(\theta_1 + \theta_2) > (\theta_2 + 1)/(\theta_2 - 1)\}.$$

Then

(i) if $\theta \in \Theta_1$, the support points are $(a - \theta_1)/\theta_2$, $(-a - \theta_1)/\theta_2$;
(ii) if $\theta \in \Theta_2$, they are -1 and u_*, where u_* is the solution of $\exp(\theta_1 + \theta_2 u) = \{2 + (u+1)\theta_2\}/\{-2 + (u+1)\theta_2\}$;
(iii) if $\theta \in \Theta_3$, they are -1 and $+1$.

This rather complicated dependence on θ of the support points may be roughly summarized by saying that the more nearly linear is the response curve in the interval $[-1, 1]$, the further towards -1 and $+1$ are the support points pushed.

Once again this analysis guides our thinking about what we might do in practice. Suppose, for instance, that we happen to know *a priori* that $\theta_2 < 1$; then $\theta \in \Theta_3$. Naturally we would take half our observations at -1 and half at $+1$, because the corresponding design measure is D_θ-optimal for all $\theta \in \Theta_3$.

Consider however, the efficiency of this design measure relative to the D_θ-optimal measure when $\theta \in \Theta_1$. A simple calculation shows that

$$J(u, \theta) = \frac{\exp(\theta_1 + \theta_2 u)}{\{1 + \exp(\theta_1 + \theta_2 u)\}^2} \begin{pmatrix} 1 & u \\ u & u^2 \end{pmatrix}.$$

Hence for the design measure μ_0 that puts probability $\frac{1}{2}$ at each of -1 and $+1$,

$$M(\mu_0, \theta) = \tfrac{1}{2}\{J(-1, \theta) + J(+1, \theta)\}$$
$$= \frac{1}{2}\begin{pmatrix} h(1,\theta) + h(-1,\theta) & h(1,\theta) - h(-1,\theta) \\ h(1,\theta) - h(-1,\theta) & h(1,\theta) + h(-1,\theta) \end{pmatrix},$$

where $h(u, \theta) = \exp(\theta_1 + \theta_2 u)/\{1 + \exp(\theta_1 + \theta_2 u)\}^2$. We have

$$\det M(\mu_0, \theta) = h(1, \theta)h(-1, \theta)$$
$$= \exp(2\theta_1)[\{1 + \exp(\theta_1 + \theta_2)\}\{1 + \exp(\theta_1 - \theta_2)\}]^{-2}.$$

Now the D_θ-optimal design measure $\mu_*(\theta)$ when $\theta \in \Theta_1$ puts probability $\frac{1}{2}$ at each of $u = (\pm a - \theta_1)/\theta_2$, and a similar calculation shows that

$$\det M\{\mu_*(\theta), \theta\} = \frac{a^2}{\theta_2^2} \exp(2a)\{1 + \exp(a)\}^{-4}.$$

We may measure the efficiency of μ_0 for estimating $\theta \in \Theta_1$ by $\det M(\mu_0, \theta)/\det M\{\mu_*(\theta), \theta\}$. The following table shows some typical efficiencies:

θ^T:	(1 3)	(1 4)	(1 5)	(1 1+a)	(2 2+a)	(3 3+a)
eff (μ_0)	.33	.10	.02	.51	.14	.03

Thus μ_0 is grossly inefficient for estimating some values of θ in Θ_1, and in the absence of any prior knowledge of θ it could be extremely wasteful in practice simply to take half our observations at -1 and half at $+1$. Of course this analysis confirms what is intuitively obvious. If θ happens to be such that the probability of response at $u = -1$ is close to 0 and that at $u = +1$ is close to 1, then observations taken at those values of u are extremely uninformative about θ.

For this example it is scarcely worth even attempting to calculate a maximin design measure because it is clear that in the absence of any prior knowledge about θ the minimum efficiency of even the maximin measure would be very low. So none of the proposals in Example 6.5 provide a practical solution to this problem; any design chosen in advance of experimentation is liable to result in wasteful experimentation. We are obliged therefore to think again and consider the possibility of sequentially constructed designs where we use information about θ obtained from previous observations to choose the next design point. This is the topic of the next chapter.

7

SEQUENTIAL DESIGNS

7.1 Objective

We recall our objective. The Fisher information matrix about θ arising from a single observation made at the value u of the vector of control variables is $J(u, \theta)$; that from n independent observations at $u_{(1)}, \ldots, u_{(n)}$ respectively is $L(\mathbf{u}_n, \theta) = \sum_{i=1}^{n} J(u_{(i)}, \theta)$, where $\mathbf{u}_n = (u_{(1)}, \ldots, u_{(n)})$. We wish to choose $\mathbf{u}_n$ to maximize a real-valued function $\phi\{L(\mathbf{u}_n, \theta)\}$ for the true parameter θ. The maximizing $\mathbf{u}_n$ usually depends on θ and since we do not know its true value we cannot achieve this objective in practice.

In the non-linear examples discussed in Chapter 6 it was possible to determine ϕ_θ-optimal design measures analytically. However typically such analytic determination will not be possible. In this case we might think of using analogues of the algorithms developed for the linear case to compute ϕ_θ-optimal design measures for a range of values of θ. However this still may not provide a practical solution to the problem. Even if it were a practical possibility to compute ϕ_θ-optimal design measures for a large enough number of values of θ, there is no guarantee that knowledge of these would help, since any design measure optimal for a particular θ might be most inefficient for other values of θ and use of a pre-determined design could result in wasteful experimentation.

We have already indicated a possible way out of this dilemma, namely to use observations already made to estimate the true θ and then to take the next observation at a value of u chosen to give a large increase of information at this estimated value. Specifically we might start with an arbitrary design $\mathbf{u}_r = (u_{(1)}, \ldots, u_{(r)})$, where $r < n$; take observations $y_1, \ldots, y_r$ at these values of u respectively, and estimate θ by $\hat{\theta}_r$, say; choose $u = u_{(r+1)}$ to maximize $\phi\{L(\mathbf{u}_r, \hat{\theta}_r) + J(u, \hat{\theta}_r)\}$; take an observation y_{r+1} at $u_{(r+1)}$; re-estimate θ from $y_1, \ldots, y_{r+1}$ by $\hat{\theta}_{r+1}$; and repeat this process with r replaced by $r+1, r+2, \ldots, n$.

SEQUENTIAL DESIGNS

There is a point worth recording at this stage. Our emphasis throughout has been on designing experiments to make Fisher's information matrix large in some sense. However if we construct a design sequentially as above, the matrix $L(\mathbf{u}_n, \theta)$ is *not* Fisher's information matrix. This is so because the observations from the $(r+1)$th are not independent. For example the distribution of $\tilde{y}_{r+1}$ depends on u_{r+1} which is determined by $y_1, \ldots, y_r$. Thus $J(u_{(r+1)}, \theta)$ is the matrix whose (i, j)th component is $E\{-\partial^2 \log p(\tilde{y}_{r+1} | y_1, \ldots, y_r, \theta) / \partial \theta_i \partial \theta_j\}$ and to calculate Fisher's information matrix for the $(r+1)$th observation we should take the expectation of this over $\tilde{y}_1, \ldots, \tilde{y}_r$. However the calculation of this latter expectation is, in general, extremely difficult. Now as far as design construction is concerned we are completely at liberty to proceed as proposed above; all that matters is that the method leads to good designs. But it is necessary to bear this point in mind when we consider the problem of making repeated-sampling inferences from data obtained from a sequentially constructed design.

7.2 An alternative method

At stage s in sequential construction, finding the u which maximizes $\phi\{L(\mathbf{u}_s, \hat{\theta}_s) + J(u, \hat{\theta}_s)\}$ may be computationally difficult. In this case, motivated by the algorithms discussed in Chapter 4, we might instead choose as the next design point that u which maximizes $F_\phi\{L(\mathbf{u}_s, \hat{\theta}_s), J(u, \hat{\theta}_s)\}$. This has been suggested by Fedorov & Malyutov (1972) and investigated in the case of D-optimality by White (1975).

When $J(u, \theta)$ has rank 1, for each u and θ, and when $\phi = \log \det$, these two approaches to the choice of $u_{(s+1)}$ are equivalent, because, writing $J(u, \hat{\theta}_s) = w_u w_u^T$, where w_u is a column vector, we have

$$F_\phi\{L(\mathbf{u}_s, \hat{\theta}_s); J(u, \hat{\theta}_s)\} = w_u^T L^{-1}(\mathbf{u}_s, \hat{\theta}_s) w_u - k;$$

see Section 5.1.1. Also, by a standard equality,

$$\det\{L(\mathbf{u}_s, \hat{\theta}_s) + w_u w_u^T\} = \det L(\mathbf{u}_s, \hat{\theta})\{1 + w_u^T L^{-1}(\mathbf{u}_s, \hat{\theta}_s) w_u\}$$

and each of the right-hand sides is maximized by the u which maximizes $w_u^T L^{-1}(\mathbf{u}_s, \hat{\theta}_s) w_u$.

However this equivalence does not hold in general. Consider, for instance, the case where $J(u, \theta)$ has rank 1 and $\phi(L) = -\operatorname{tr} L^{-1}$. Then the method of Section 7.1 leads to choosing u to maximize

$$\frac{w_u^T L^{-2}(\mathbf{u}_s, \hat{\theta}_s) w_u}{1 + w_u^T L^{-1}(\mathbf{u}_s, \hat{\theta}_s) w_u},$$

while that of this section involves maximizing simply the numerator of this expression.

Whether or not the two methods of choosing the next design point are equivalent, the basic idea underlying both is the same, since $F_\phi(L, J)$ can be regarded as a first order approximation to $\phi(L + J) - \phi(L)$. Any other approximation to this latter expression might be used instead without affecting greatly the resulting design.

7.3 Convergence considerations

The heuristic asymptotic justification of sequential construction is fairly obvious. One expects that, as $n \to \infty$, $\hat{\theta}_n$ will tend with probability 1 to the true parameter, θ_*, say. If this is so, then at stage s of the sequential procedure, when s is very large, we will be choosing the next design point effectively as if we know the true parameter. Then by the kind of convergence arguments we had for algorithms the information matrix of the design measure corresponding to the sequentially constructed n-observation design $\mathbf{u}_n = (u_{(1)}, \ldots, u_{(n)})$ will converge to the information matrix of a ϕ_{θ_*}-optimal design measure.

Conversion of this heuristic argument into a formal general proof appears to be difficult and such results as have been proved tend to beg the question; see White (1975). However formal proof is possible in particular cases. White (1975) gives one example. Ford and Silvey (1980) give another and since this contains certain points of interest we shall now discuss it briefly.

It is the example previously considered in Example 6.4 where $E(\tilde{y}|u, \theta) = \theta_1 u + \theta_2 u^2$, the design space is $[-1, 1]$, and we are interested in designing to estimate $g(\theta) = -\theta_1/(2\theta_2)$. As before, for a design measure μ, our criterion function ϕ is given by

$$\phi\{M(\mu), \theta\} = -c_g^T M^-(\mu) c_g,$$

where $c_g^T = (1 \quad 2g)$; see Section 6.4.1. It is not difficult to show that the ϕ_θ-optimal design measure has information matrix

$$M_\theta = \begin{pmatrix} 1 & k(\theta) \\ k(\theta) & 1 \end{pmatrix},$$

where $k(\theta) = 2g(\theta)$ if $|g(\theta)| \leq \frac{1}{2}$ and $k(\theta) = \{2g(\theta)\}^{-1}$ if $|g(\theta)| \geq \frac{1}{2}$. We observe again that if $|g(\theta)| = \frac{1}{2}$, then M_θ is singular.

If we assume normal errors of unit variance, Fisher's information matrix for a single observation on y at the value u of the control variable is

$$J(u) = \begin{pmatrix} u^2 & u^3 \\ u^3 & u^4 \end{pmatrix}.$$

Suppose that we construct a design sequentially, adopting the procedure of Section 7.2. Specifically, we start by taking one observation at each of $u = \pm 1$. Thereafter at stage s we choose the next design point $u = u_{(s+1)}$ to maximize $F_{\phi_s}\{L(\mathbf{u}_s), J(u)\}$, where $\phi_s = \phi(\cdot, \hat{\theta}_s)$, $L(\mathbf{u}_s) = \sum_{i=1}^{s} J(u_{(i)})$ and $\hat{\theta}_s$ is the maximum likelihood estimator of θ obtained from the first s observations. What are the properties of a design so constructed?

Ford and Silvey (1980) have investigated this problem both theoretically and by means of a simulation study. Leaving aside for the moment inference questions, the main conclusions to emerge were as follows.

(i) The method of design construction forces all observations to be taken at either $u = +1$ or $u = -1$. It follows that if s_n is the number of the first n observations to be taken at $u = -1$, then

$$L(\mathbf{u}_n) = \begin{pmatrix} n & n - 2s_n \\ n - 2s_n & n \end{pmatrix}.$$

As observed earlier $L(\mathbf{u}_n)$ is *not* Fisher's information matrix; s_n is a random variable and Fisher's information matrix is obtained from $L(\mathbf{u}_n)$ by replacing s_n by its expected value.

(ii) If θ_* is the true value of θ, then $(1/n)L(\mathbf{u}_n) \to M_{\theta_*}$, almost certainly, as $n \to \infty$, and this holds whether or not M_{θ_*} is singular. In this sense the method of design construction is asymptotically efficient.

(iii) The maximum-likelihood estimator $\hat{\theta}_n$ is consistent even in the singular case. There is a point of some interest here so far as general theory is concerned. As we have noted earlier, such general theorems as have been proved tend to beg the question and, when applied to our problem might read 'if $(1/n)L(\mathbf{u}_n)$ tends to a non-singular matrix as $n \to \infty$, then $\hat{\theta}_n$ is consistent'. Hence the way in which consistency is achieved in the singular cases, that is, when $g(\theta) = \pm\frac{1}{2}$ is worth noting. When $g(\theta) = \pm\frac{1}{2}$, $s_n/n \xrightarrow{\text{a.c.}} 0$ or 1 according to the sign of $g(\theta)$, and this results in singularity in the limit of $(1/n)L(\mathbf{u}_n)$. However in either case both s_n and $n - s_n$ tend to infinity as $n \to \infty$ and this results in consistency of $\hat{\theta}_n$. Consistency and singularity of the limiting matrix may occur side by side in other problems for similar reasons.

(iv) For each pair of true parameter values adopted in the simulation study, namely $\theta = (1, 4)$, $(1, 2)$, $(1, 1)$ and $(4, 4)$, the empirical distribution of s_n was quite highly concentrated even for n as small as 25; that is, the achieved design varied very little over 1000 sequentially constructed designs. Moreover the value of s_n/n rapidly approached, in most cases, the weight placed at $u = -1$ by the ϕ_θ-optimal design measure. Thus the method of design construction is not only asymptotically efficient, but efficient also for relatively small designs.

It would be interesting to know how generally the conclusions of this particular study apply: efficiency, stability, consistency. But this remains an open question, one which seems likely to require further theoretical and empirical investigation.

7.4 *Inference from sequentially constructed designs*

In constructing sequential designs we found it expedient to work at stage s with the matrix $\sum_{i=1}^{s} J(u_{(i)}, \theta)$ rather than with Fisher's information matrix. As we have stated earlier this raises no question of principle; all that matters is that the method is effective in practice and we have just demonstrated for a particular example that it is indeed effective. However, having constructed a design sequentially, we will usually wish to make inferences about θ and questions arise about how sequential construction affects this activity.

If either a Bayesian or likelihood approach to inferences is adopted there is no problem. The likelihood function arising from a sequentially constructed design is just the same as if the design had not been constructed sequentially and we had merely taken independent observations on y at the pre-determined design points $u_{(1)}, \ldots, u_{(n)}$.

However if a repeated-sampling approach is adopted, the situation is not quite so clear-cut. The repeated-sampling properties of sequential designs are quite different from those of pre-determined designs; the design achieved varies from occasion to occasion in the former, but not in the latter.

Suppose, in particular, that we wish to construct a confidence region for θ. For a pre-determined design $\mathbf{u}_n = (u_{(1)}, \ldots, u_{(n)})$, with independent observations at these design points, $L(\mathbf{u}_n, \theta) = \sum_{i=1}^{n} J(u^{(i)}, \theta)$ is Fisher's information matrix. Given sufficient regularity, standard large-sample theory applies and, for large n, the maximum-likelihood estimator $\hat{\theta}_n$ of θ is approximately normally distributed with mean θ and variance matrix $L^{-1}(\mathbf{u}_n, \theta)$. From this, confidence regions for θ of given confidence coefficient can be constructed. But suppose that the design $\mathbf{u}_n$ has been achieved by sequential construction. Is $L^{-1}(\mathbf{u}_n, \theta)$ still an adequate approximation to the variance matrix of $\hat{\theta}_n$? Should we use instead the inverse of Fisher's information matrix? Is $\hat{\theta}_n$ approximately normally distributed?

The probability calculus involved in determining the repeated-sampling properties of an estimator obtained from a sequentially constructed design is quite intractable. Even asymptotic theory is not straightforward because of the dependence among observations. Empirical investigation of questions like those posed above therefore seems to be necessary. Ford and Silvey (1980) looked at some of these questions in the simulation study of the example discussed in Section 7.3 and for what it is worth we consider now the main conclusions.

Recall the example. We had $E(\tilde{y}|u, \theta) = \theta_1 u + \theta_2 u^2$, the design space was $[-1, 1]$ and we were interested in designing to estimate $g(\theta) = -\theta_1/(2\theta_2)$. For various pairs of parameter values, 1000 sequential designs each of 100 observations were constructed using generated $N(0, 1)$ errors. Various statistics were calculated after $n = 25, 50, 75, 100$ observations. Regarding repeated-sampling properties and inferences the following main conclusions emerged.

(i) In non-singular cases, that is, when θ is such that the ϕ_θ-optimal design measure is non-singular, the empirical sampling distribution of $\hat{g}_n$ rapidly approached normality and was centred very close to the true value of g, even for n as small as 25. In singular cases the approach to normality seemed slower, the empirical variance of $\hat{g}_n$, for each given n, was larger but its distribution was still centred close to the true g, even for relatively small n.

(ii) If $L^{-1}(\mathbf{u}_n)$ was used as an approximation to the variance matrix of $\hat{\theta}_n$, normality of the distribution of $\hat{g}_n$ was assumed and an approximate 95 % confidence interval for g constructed accordingly, the empirical percentage coverage of the true g was, for each n, not far from 95 %, certainly in the non-singular cases. In the singular cases investigated, the empirical percentage coverage departed rather more from 95 %, but this appeared to be unrelated to the fact that the design was sequential but due rather to the fact that θ_2 was not particularly well estimated.

(iii) Using an approximation to the inverse of Fisher's information matrix rather than $L^{-1}(\mathbf{u}_n)$ to approximate var $(\hat{\theta}_n)$ gave similar results. But this is not surprising for this particular example because of the stability of the achieved design referred to in Section 7.3.

(iv) The empirical conditional distribution of $\hat{g}_n$, given the achieved design, that is, given s_n, the number of observations at $u = -1$, was of some interest. This appeared to be centred on the value of g corresponding to values of θ for which the ϕ_θ-optimal design measure puts weight s_n/n at $u = -1$; and of course the empirical conditional variance of $\hat{g}_n$, given s_n, was much smaller than its unconditional variance. Hence inferences about g made conditionally on s_n could be quite misleading.

The extent to which similar conclusions apply to other examples is, of course, open to question. They may not exhibit the same stability in achieved design and then the question of whether it is better to use $L^{-1}(\mathbf{u}_n, \hat{\theta}_n)$ or the inverse of Fisher's information matrix as an approximation to var $(\hat{\theta}_n)$ has considerably more content. Clearly further empirical investigation of this kind of problem is necessary before any general recommendations about repeated-sampling inferences from data obtained from sequential designs can be made. For a fairly recent more comprehensive review of sequential design methods, see Chernoff (1975).

Appendix 1
CONCAVITY RESULTS

We here establish the concavity of criterion functions on which we have laid emphasis. For other results on concavity (convexity) see Kiefer (1974).

Let A be a $k \times s$ matrix of rank s and let $\mathcal{M}_A$ be the set of non-negative definite $k \times k$ matrices M such that $A = MY$ for some Y, or, equivalently, such that $Mz = 0$ implies $A^T z = 0$. It is not difficult to verify that $\mathcal{M}_A$ is a convex set.

For $M \in \mathcal{M}_A$ define the matrix-valued function f by

$$f(M) = A^T M^- A,$$

where M^- is any generalized inverse of M, that is, any matrix such that $MM^- M = M$. The function $f(M)$ does not depend on which g-inverse is chosen, since when $A = MY$

$$A^T M^- A = Y^T MM^- MY = Y^T MY = Y^T A.$$

Also $f(M)$ is a positive definite $s \times s$ matrix.

Let $\geqslant$ denote the usual partial ordering on the symmetric $s \times s$ matrices; that is, $S_1 \geqslant S_2$ iff $S_1 - S_2$ is non-negative definite. Then:

A.1 $f(M)$ is convex in the sense that $f\{\lambda M_1 + (1 - \lambda)M_2\} \leqslant \lambda f(M_1) + (1 - \lambda)f(M_2)$ for $0 \leqslant \lambda \leqslant 1$, $M_1, M_2 \in \mathcal{M}_A$.

A.2 $\{f(M)\}^{-1}$ is concave in a similar sense.

Proof of A.1. For convenience we use the Moore–Penrose g-inverse which we shall denote by M^+.

Let R be any $k \times s$ matrix. We have, when $M \in \mathcal{M}_A$,

$$\{(M^+)^{1/2}A - M^{1/2}R\}^T\{(M^+)^{1/2}A - M^{1/2}R\} = A^T M^+ A - A^T R - R^T A + R^T MR,$$

since when $M \in \mathcal{M}_A$, $A^T(M^+)^{1/2}M^{1/2} = A^T$, as is readily verified.

Hence

$$A^T M^- A = A^T M^+ A \geqslant A^T R + R^T A - R^T M R,$$

with equality when $(M^+)^{1/2} A = M^{1/2} R$. This means that on $\mathcal{M}_A$, $f(M)$ is the upper envelope of a family of linear functions and so it is convex.

Proof of A.2. It is clearly sufficient to prove the result for $A^T = (I_s \ \ 0)$. The general result then follows by changing bases in R^s and R^k.

Partition M into

$$\begin{pmatrix} M_{11} & M_{12} \\ M_{12}^T & M_{22} \end{pmatrix}$$

where M_{11} is $s \times s$; if $A = MY$ partition Y into

$$\begin{pmatrix} Y_1 \\ Y_2 \end{pmatrix}$$

conformably. Now $A^T = (I_s \ \ 0)$ and $A = MY$ imply

$$M_{11} Y_1 + M_{12} Y_2 = I_s$$

$$M_{12}^T Y_1 + M_{22} Y_2 = 0.$$

Let C be any solution of the equation $CM_{22} = M_{12}$. Multiplying the second of the above equations by C and subtracting gives

$$(M_{11} - CM_{22}C^T) Y_1 = I_s$$

which implies that each matrix on the left-hand side is non-singular and $Y_1 = (M_{11} - CM_{22}C^T)^{-1}$. Now when $A^T = (I_s \ \ 0)$

$$f(M) = Y^T A = Y_1^T = (M_{11} - CM_{22}C^T)^{-1}$$

and

$$\{f(M)\}^{-1} = M_{11} - CM_{22}C^T.$$

As before, let R be any $s \times (k-s)$ matrix. Then

$$M_{11} - CM_{22}C^T \leqslant M_{11} - CM_{22}C^T + (C+R)M_{22}(C+R)^T$$

with equality when $C = -R$. Now the right-hand side of this equation is

$$M_{11} + M_{12} R^T + R M_{12}^T + R M_{22} R^T.$$

Hence $\{f(M)\}^{-1}$ is the lower envelope of a family of linear functions and so is concave.

Since 'trace' is an increasing linear function on the non-negative definite matrices, it follows easily from Statement A.1 that tr $(A^T M^- A)$ is convex on $\mathcal{M}_A$; or equivalently that $-\text{tr}(A^T M^- A)$ *is concave on* $\mathcal{M}_A$.

Further 'log det' is an increasing concave function on the positive definite matrices; see, for instance, Fedorov (1972, p. 20). It follows from Statement A.2 that $-\log \det (A^T M^- A)$ *is concave on* $\mathcal{M}_A$.

The idea of expressing $f(M)$ and $\{f(M)\}^{-1}$ as envelopes of families of linear functions is due to Sibson (personal communication).

Appendix 2
CARATHÉODORY'S THEOREM

The *convex hull*, written conv S, of a set S in n-dimensional Euclidean space R^n is the set of all finite convex combinations of elements of S, that is, the set of all vectors of the form $\lambda_1 x_1 + \ldots + \lambda_m x_m$ with each $\lambda_i > 0, \sum \lambda_i = 1$, each $x_i \in S$ and m any positive integer. This is the smallest convex set containing S.

Carathéodory's Theorem shows that conv S is generated by convex combinations of at most $n + 1$ elements of S. We prove the theorem by considering a convex cone in R^{n+1}.

The *convex cone* generated by a set S in R^n is the set of all positive finite linear combinations of elements of S, that is, the set of all vectors of the form $\lambda_1 x_1 + \ldots + \lambda_m x_m$ with each $\lambda_i > 0$, each $x_i \in S$, and m any positive integer. Here of course it is not required that $\sum \lambda_i = 1$.

Given a set S in R^n, let S' be the subset of R^{n+1} consisting of vectors of the form $(1 \quad x)$ with $x \in S$; and let K be the convex cone generated by S'. Let C be the set of vectors c in R^n such that $(1 \quad c) \in K$. Then it is easily seen that $C = \text{conv } S$.

Theorem (Carathéodory). *Every element C in conv S can be expressed as a convex combination of at most $n + 1$ elements of S. If c is in the boundary of conv S, $n + 1$ can be replaced by n.*

Proof. Let S' and K be as above and let $y \in K$ so that

$$y = \lambda_1 y_1 + \ldots + \lambda_m y_m$$

where each $\lambda_i > 0$ and each $y_i \in S'$. Suppose that the y_i's are not linearly independent; they certainly will not be if $m > n + 1$. Then there exist $\mu_1, \ldots, \mu_m$ not all zero and such that

$$\mu_1 y_1 + \ldots + \mu_m y_m = 0.$$

Since the first component of each y_i is 1 we have $\mu_1 + \ldots + \mu_m = 0$ and so at least one μ_i is positive. Let λ be the largest number such that $\lambda \mu_i \leq \lambda_i, i = 1, \ldots, m$; λ is finite since at least one μ_i is

CARATHÉODORY'S THEOREM

positive. Now let $\lambda'_i = \lambda_i - \lambda\mu_i$. Then

$$\lambda'_1 y_1 + \ldots + \lambda'_m y_m = \lambda_1 y_1 + \ldots + \lambda_m y_m - \lambda(\mu_1 y_1 + \ldots + \mu_m y_m) = y$$

and at least one $\lambda'_i = 0$. We have thus expressed y as a positive linear combination of fewer than m elements of S'.

This argument can clearly be repeated until y has been expressed as a positive linear combination of at most $n + 1$ elements of S', since more than $n + 1$ elements are necessarily linearly dependent.

In particular all elements y of the form (1 c) can be expressed in this way and the first part of the theorem follows immediately.

Now suppose that $y \in K$ and

$$y = \lambda_1 y_1 + \ldots + \lambda_{n+1} y_{n+1}$$

where each $\lambda_i > 0$ and the y_i's are linearly independent. Any element z of R^{n+1} can be expressed as

$$z = \mu_1 y_1 + \ldots + \mu_{n+1} y_{n+1}.$$

Clearly since each $\lambda_i > 0$, then for small enough $\varepsilon > 0$, $y + \varepsilon z \in K$. Therefore y is in the interior of K. It follows that any boundary point of K can be expressed as a positive linear combination of at most n linearly independent elements of S' and the second part of the theorem is an immediate consequence.

In this monograph we have been concerned with the convex hull $\mathcal{M}$ of a set S in $R^{(1/2)k(k+1)}$, where

$$S = \{xx^T : x \in \mathcal{X}\},$$

$\mathcal{X}$ being a compact subset of R^k. We defined $\mathcal{M}$ in the following way:

Let η be a probability distribution on the Borel sets of $\mathcal{X}$, H the set of *all* η and $\tilde{x}$ a random vector with distribution η. Then

$$\mathcal{M} = \{M = E(\tilde{x}\tilde{x}^T) : \eta \in \mathrm{H}\}.$$

It may not be immediately obvious that $\mathcal{M} = \mathrm{conv}\, S$, but this does not really matter since we might as well have defined H to be the class of all discrete probability distributions on $\mathcal{X}$ with finite support. With H so defined, the above definition of $\mathcal{M}$ coincides with that of conv S, though it is expressed in terms of expectations of random vectors rather than convex combinations of elements of S.

Appendix 3
DIFFERENTIABILITY

In this section we recall various results on differentiability of real-valued functions defined on R^n and indicate their relevance to our study. These are all paraphrases of results to be found in Rockafeller (1970) and page references quoted (R. p. 000) are to that book. It suits our convenience to consider functions that are defined on all of R^n, possibly taking the values $\pm\infty$ on part of it: R. p. 23.

1 *Differentiability:* R. p. 241.

Let f be a function from R^n to $[-\infty, \infty]$ and $x = (x_1, \ldots, x_n)$ a point where f is finite. Then f is differentiable at x iff there exists x^*, necessarily unique, with the property that, for all y,

$$f(y) = f(x) + \langle x^*, y - x \rangle + o(\|y - x\|).$$

Here $\langle \cdot, \cdot \rangle$ denotes the Euclidean inner product, $\langle x, y \rangle = \sum x_i y_i$, and $\|\ \|$ Euclidean distance.

Such an x^*, if it exists, is called the gradient of f at x; it is often denoted by $\nabla f(x)$, and

$$\nabla f(x) = \left(\frac{\partial f(x)}{\partial x_1}, \ldots, \frac{\partial f(x)}{\partial x_n} \right).$$

2 *Directional derivatives*

The Gâteaux derivative $G_f(x, y)$ of f at x in the direction of y is defined by

$$G_f(x, y) = \lim_{\varepsilon \to 0^+} \frac{1}{\varepsilon} \{ f(x + \varepsilon y) - f(x) \}.$$

(The Gâteaux derivative is denoted $f'(x, y)$ by Rockafeller.)

DIFFERENTIABILITY

If f is differentiable at x, then $G_f(x, y)$ exists and

$$G_f(x, y) = \langle \nabla f(x), y \rangle = \sum y_i \frac{\partial f(x)}{\partial x_i}.$$

We note that

$$G_f(x, y - x) = \langle \nabla f(x), y - x \rangle$$

may, in virtue of the definition of $\nabla f(x)$, be regarded as a first-order approximation to $f(y) - f(x)$, that is

$$f(y) - f(x) = G_f(x, y - x) + o(\|y - x\|),$$

when f is differentiable at x. We have called $G_f(x, y - x)$ the Fréchet derivative of f at x in the direction of y, and denoted it by $F_f(x, y)$.

3 Concave and convex functions

3.1 R. p. 25. A function f from R^n to $[-\infty, \infty)$ is concave if

$$f\{(1 - \lambda)x + \lambda y\} \geqslant (1 - \lambda)f(x) + \lambda f(y), \qquad 0 \leqslant \lambda \leqslant 1.$$

Note that f may take the value $-\infty$.

A concave function is *proper* if it is defined and finite on a convex subset C of R^n and takes the value $-\infty$ outside C. The effective domain of f is then C, written dom f.

A very important property of concave functions is the following:

3.2 R. p. 213. If f is a concave function and x a point where f is finite, then $G_f(x, y)$ exists for all y; this is so whether or not f is differentiable at x.

4 Differentiability of concave functions

4.1 R. p. 244. If f is a concave function and x a point where f is finite, then f is differentiable at x iff $G_f(x, y)$ is linear in its second argument, that is,

$$G_f(x, \sum a_i y_{(i)}) = \sum a_i G_f(x, y_{(i)}),$$

for all real a_i and $y_{(i)} \in R^n$.

4.2 R. p. 243. If f is a concave function which is finite and differentiable at a point x, then f is proper and $x \in \text{int}(\text{dom} f)$.

5 Functions occurring in optimal design

In optimal design theory we are concerned with functions of non-negative definite $k \times k$ matrices. The symmetric $k \times k$ matrices may be regarded as elements of $R^{(1/2)k(k+1)}$ or as elements of $R^{k \times k}$. While there are certain advantages in the latter view, particularly in the ease of definition of inner product, it suits our convenience to view them in the former way. When we do so, the non-negative definite matrices form a convex cone $\mathcal{N}$ in $R^{(1/2)k(k+1)}$ and the interior of this cone consists of the positive definite matrices.

The functions ϕ of interest to us are concave on $\mathcal{N}$. Since our interest in them is confined to their behaviour on $\mathcal{N}$, it is natural so to define them that they take the value $-\infty$ off $\mathcal{N}$, even though they may be 'naturally' defined for some symmetric matrices not in $\mathcal{N}$. Thus while log det is 'naturally' defined and finite for some symmetric matrices outside $\mathcal{N}$, we define a function ϕ on *all* the symmetric $k \times k$ matrices by

$$\phi(N) = \log \det N, \quad \text{if } N \text{ is positive definite}$$
$$= -\infty, \quad \text{otherwise}.$$

This is quite a reasonable thing to do because all functions of concern to us are typically finite on int $\mathcal{N}$ and 'naturally' take the value $-\infty$ for at least some points in the boundary of $\mathcal{N}$. Thus log det naturally takes the value $-\infty$ at all points in the boundary of $\mathcal{N}$, where matrices are singular. On the other hand if $\phi(N) = -\text{tr } A^T N^- A$ when $A = NY$ for some Y, ϕ is 'naturally' finite both on int $\mathcal{N}$ and at some points on its boundary, but 'naturally' $-\infty$ at other such points, namely at those semi-definite N such that there is no Y for which $A = NY$.

The effective domain of all functions of interest to us is a convex subset of $\mathcal{N}$ whose interior coincides with int $\mathcal{N}$. By virtue of Section 4.2 it is not surprising to find that these functions are not differentiable at points in the boundary of $\mathcal{N}$ even when they are finite at such points.

When a function ϕ, defined on the $k \times k$ symmetric matrices is differentiable at a point N, $\nabla \phi(N)$ is really an inconvenient

DIFFERENTIABILITY

entity with which to work. However if ϕ is concave or convex we can avoid $\nabla\phi(N)$ altogether by making use of Section 4.1. The Gâteaux derivative $G_\phi(N, \cdot)$ exists when $\phi(N)$ is finite; typically it is relatively easy to calculate from first principles; and the differentiability status of ϕ at N can be determined by noting whether or not $G_\phi(N, \cdot)$ is a linear function. We have exploited this notion in order to simplify the presentation of the general theorems of optimal linear regression design.

Appendix 4
LAGRANGIAN THEORY: DUALITY

1   In this monograph we have in the main exploited differential calculus to provide the important theorems of optimal linear regression design and we have only mentioned in passing the use of duality theory for this purpose. However recent results, particularly those of Pukelsheim (1980), suggest that the latter may be a more powerful tool and because of this we here give a very brief and superficial account of the ideas involved, together with Sibson's (1972) proof of the duality between the minimum-ellipsoid problem and the D-optimal design problem. For a deeper discussion of duality theory see Rockafeller (1970).

Let f be a real-valued function defined on some subset X of R^n and consider the problem $P(b)$: maximize f over X subject to the constraints $g_1(x) = b_1, \ldots, g_m(x) = b_m$, where $g_1, \ldots, g_m$ are real-valued functions on R^n and $b_1, \ldots, b_m$ are real numbers. We shall write these constraints as $g(x) = b$, where g is a vector-valued function and b is an m-vector.

The classical method of solving this problem is to introduce Lagrange multipliers $\lambda_1, \ldots, \lambda_m$ and to consider stationary values of the Lagrangian

$$f(x) + \sum_{i=1}^{m} \lambda_i \{b_i - g_i(x)\} = f(x) + \lambda^T \{b - g(x)\}.$$

In certain circumstances, $\bar{x}$, a value of x which solves $P(b)$ emerges, for suitable λ, as one of these stationary values.

This method depends on the validity of the Weak Lagrangian Principle which states: there exists a vector λ such that the Lagrangian is stationary at any $\bar{x}$ which solves $P(b)$.

For certain problems a stronger principle is valid, the Strong Lagrangian Principle which states: there exists a vector λ such that the Lagrangian is *maximal* at any $\bar{x}$ which solves $P(b)$. Sufficient conditions for the validity of this Principle are given by Whittle (1971).

When the Strong Principle is valid there is more that can be said about the appropriate λ. It is such as to minimize

$$\max_{x \in X} [f(x) + \lambda^T\{b - g(x)\}].$$

Thus corresponding to the *primal* maximization problem $P(b)$, there is a minimization problem called the *dual* problem. These primal and dual problems share a common extreme value. Validity of the Strong Principle implies that we can solve $P(b)$ by maximizing the Lagrangian over X without constraints for each λ and then finding the λ which minimizes the resulting maximum.

A problem that seems different at first sight is $P_1(b)$: maximize $f(x)$ over X subject to $g_i(x) \leqslant b_i$, $i = 1, \ldots, m$; that is, a problem involving inequality rather than equality constraints. However it is not essentially different because we can introduce *slack* variables $s_1, \ldots, s_m$ and write $P_1(b)$ as: maximize $f(x)$ for $x \in X$, $s \geqslant 0$, $g(x) + s = b$. The problem $P_1(b)$ now has exactly the same form as $P(b)$ with (x, s) replacing x, $X \times R_+^m$ replacing X, where R_+^m is the positive octant of R^m, and $g^*(x, s) = g(x) + s$ replacing $g(x)$.

The Strong Lagrangian Principle now asserts: there exists a vector λ such the Lagrangian

$$f(x) + \lambda^T\{b - g(x) - s\}$$

is maximal at any $(\bar{x}, \bar{s})$ which solves $P_1(b)$. When this is so, the dual problem to $P_1(b)$ is to find λ to minimize

$$\max_{x \in X, s \in R_+^m} [f(x) + \lambda^T\{b - g(x) - s\}].$$

We may then solve $P_1(b)$ by maximizing the Lagrangian for each fixed λ and then finding the λ which minimizes this maximum.

2 We turn now to the application of these ideas to the D-optimum design problem. Geometrical considerations suggest a primal problem of which it is the dual. This primal problem is the so-called minimum-ellipsoid problem which is: given a compact set $\mathcal{X}$ in R^k, find the ellipsoid centred on the origin, containing $\mathcal{X}$ and having minimum content.

We note that if A is a positive definite $k \times k$ matrix and u is a k-vector then $\{u : u^T A u = k\}$ is the surface of an ellipsoid centred on the origin and any such ellipsoid can be described in this way;

here we choose k on the right-hand side rather than 1 for subsequent algebraic simplicity. The content of this ellipsoid is proportional to $(\det A)^{-1/2}$ and the ellipsoid contains $\mathscr{X}$ iff $x^T A x \leqslant k$ for all $x \in \mathscr{X}$. Hence an analytic statement of the minimum ellipsoid problem for $\mathscr{X}$ is:

Find a positive definite $k \times k$ matrix A to maximize $\log \det A$ subject to $x^T A x \leqslant k$ for all $x \in \mathscr{X}$.

We work with $\log \det A$ rather than $\det A$ simply to ensure concavity of the objective function and consequent easy verification of the validity of the Strong Lagrangian Principle. The theorem proved by Sibson (1972) is the following.

Theorem. *If $\mathscr{X}$ is a compact subset of R^k, the D-optimal design problem with $\mathscr{X}$ as design space is the dual of the minimum-ellipsoid problem for $\mathscr{X}$.*

Proof. Suppose first that $\mathscr{X}$ is finite, consisting of the vectors $x_{(1)}, \ldots, x_{(J)}$. Then we have to find positive definite A to maximize $\log \det A$ subject to the inequality constraints

$$x_{(j)}^T A x_{(j)} \leqslant k, \quad j = 1, \ldots, J.$$

Now $\log \det A$ is a concave function on the convex cone consisting of the positive definite $k \times k$ matrices and the constraining inequalities are linear in A. The Strong Lagrangian Principle is therefore valid for this problem and, introducing a J-vector λ of Lagrange multipliers and a positive vector s of slack variables, we know that there exists a λ such that the Lagrangian

$$\log \det A + \sum_j \lambda_j (k - x_{(j)}^T A x_{(j)} - s_j)$$

is maximal at $(\bar{A}, \bar{s})$, any solution to the minimum-ellipsoid problem; and that the appropriate λ is such as to minimize the maximum with respect to A and s of the Lagrangian.

All the elements of λ must be non-negative for otherwise the Lagrangian could be made arbitrarily large and we know that for the appropriate λ it is bounded above. The s_j corresponding to any positive λ_j must be zero for a maximum. Hence the maximizing problem reduces to: find positive definite A to maximize

$$\log \det A - \sum \lambda_j x_{(j)}^T A x_{(j)} + k \sum \lambda_j.$$

This expression equals

$$\log \det A - \text{tr}\{AM(\lambda)\} + k\sum \lambda_j,$$

where $M(\lambda) = \sum \lambda_j x_{(j)} x_{(j)}^T$.

Now λ must be such that $M(\lambda)$ is positive definite since otherwise the above expression could once again be made arbitrarily large; and when $M(\lambda)$ is positive definite the expression is uniquely maximized at $\bar{A} = M^{-1}(\lambda)$, its maximum value being

$$-\log \det M(\lambda) + k(\sum \lambda_j - 1).$$

The appropriate λ minimizes this expression. Suppose that $\sum \lambda_j \neq 1$ and let $\eta_j = \lambda_j / \sum \lambda_j$, all j. Then

$$\begin{aligned}-\log \det M(\eta) + k(\sum \eta_j - 1) &= -\log \det M(\eta)\\ &= -\log \det M(\lambda) + k \log \sum \lambda_j\\ &< -\log \det M(\lambda) + k(\sum \lambda_j - 1),\end{aligned}$$

since $\log z \leq z - 1$ with equality only when $z = 1$. Hence the appropriate λ must be such that $\sum \lambda_j = 1$, and, as noted earlier, each $\lambda_j \geq 0$.

The dual of the minimum-ellipsoid problem therefore reduces to: find non-negative λ_j with $\sum \lambda_j = 1$ to minimize $-\log \det M(\lambda)$, and this of course is just the D-optimum design problem.

Now suppose that $\mathcal{X}$ is not finite. By Section 3.2.3 there exists a finite subset of $\mathcal{X}$, say $\bar{\mathcal{X}}$, for which the D-optimum design problem is the same as it is for the whole of $\mathcal{X}$. Let $\bar{\lambda}$ be a D-optimum design measure for $\bar{\mathcal{X}}$ and so for $\mathcal{X}$. Suppose that the minimum ellipsoid for $\bar{\mathcal{X}}$ does not contain the whole of $\mathcal{X}$. Then by adding a suitable point to $\bar{\mathcal{X}}$ we could find a finite subset of $\mathcal{X}$ for which the minimum-ellipsoid would be larger than that for $\bar{\mathcal{X}}$. By the duality established for finite subsets this would imply the existence of a design measure λ on $\mathcal{X}$ for which $-\log \det M(\lambda)$ was smaller than $-\log \det M(\bar{\lambda})$; a contradiction. This completes the proof.

Note that this duality theorem implies that a design measure λ_* is D-optimal iff $x^T M^{-1}(\lambda_*) x \leq k$ for all $x \in \mathcal{X}$ and that there is equality at the support points of λ_*.

The argument we have presented is, of course, very specific to the D-optimal design problem. Pukelsheim (1980) uses duality theory in a deeper and more general manner to obtain similar duality theorems for a wide class of design criterion functions.

REFERENCES

ASH, A. and HEDAYAT, A. (1978). An introduction to design optimality with an overview of the literature. *Commun. Statist. A7*, **14**, 1295–1325.

ATWOOD, C.L. (1969). Optimal and efficient designs of experiments. *Ann. Math. Statist.*, **40**, 1570–1602.

ATWOOD, C.L. (1973). Sequences converging to D-optimal designs of experiments. *Ann. Statist.*, **1**, 342–352.

ATWOOD, C.L. (1976). Convergent design sequences, for sufficiently regular optimality criteria. *Ann. Statist.*, **4**, 1124–1138.

BANDEMER, H. (1979). Problems in foundation and use of optimal experimental designs in regression models. Preprint.

BOX, G.E.P. and DRAPER, N.R. (1959). A basis for the selection of a response surface design. *J. Amer. Statist.*, **54**, 622–654.

BOX, G.E.P. and HUNTER, W.G. (1965a). Sequential design of experiments for non linear models. *Proc. IBM Sc. Comp. Symp. 1963*, IBM, New York.

BOX, G.E.P. and HUNTER, W.G. (1965b). The experimental study of physical mechanisms. *Technometrics*, **7**, 23–42.

BOX, G.E.P. and LUCAS, H.L. (1959). Design of experiments in non-linear situations. *Biometrika*, **46**, 77–90.

BOX, G.E.P. and WILSON, K.B. (1951). On the experimental attainment of optimum conditions. *J.R. Statist. Soc. B.*, **13**, 1–45.

BROOKS, R.J. (1972). A decision theory approach to optimal regression designs. *Biometrika*, **59**, 563–571.

CHERNOFF, H. (1953). Locally optimal designs for estimating parameters. *Ann. Math. Statist.*, **24**, 586–602.

CHERNOFF, H. (1975). *Approaches in sequential design of experiments*. A survey of statistical design and linear models (J.N. Srivastava, Ed.). New York: American Elsevier, 67–90.

ELFVING, G. (1952). Optimum allocation in linear regression theory. *Ann. Math. Statist.*, **23**, 255–262.

ELFVING, G. (1955). Geometric allocation theory. *Skand. Aktuarietidskr.*, **37**, 170–190.

ELFVING, G. (1959). Design of linear experiments. *Cramer Festschrift Volume*. New York: Wiley, 58–74.

FEDOROV, V.V. (1972). *Theory of Optimal Experiments*. New York: Academic Press.

FEDOROV, V.V. and MALYUTOV, M.B. (1972). Optimal design in regression experiments. *Math. Operationsforsch. Statist.*, **14**, 237–324.

FORD, I. (1976). Ph.D. Thesis, University of Glasgow.

REFERENCES

FORD, I. and SILVEY, S.D. (1980). A sequentially constructed design for estimating a non-linear parametric function. *Biometrika*, **67**, 000–000.

de la GARZA, A. (1954). Spacing of information in polynomial regression. *Ann. Math. Statist.*, **25**, 123–130.

de la GARZA, A. (1956). Quadratic extrapolation and a related test of hypothesis. *J. Amer. Statist. Assoc.*, **27**, 644–649.

GUEST, P.G. (1958). The spacing of observations in polynomial regression. *Ann. Math. Statist.*, **29**, 294–299.

HILL, W.J. and HUNTER, W.G. (1974). Design of experiments for subsets of parameters. *Technometrics*, **16**, 425–434.

HOEL, P.G. (1961). Asymptotic efficiency in polynomial estimation. *Ann. Math. Statist.*, **32**, 1042–1047.

HOEL, P.G. (1965). Optimum designs for polynomial extrapolation. *Ann. Math. Statist.*, **36**, 1483–1493.

HUNTER, W.G., HILL, W.J. and HENSON, T.L. (1969). Designing experiments for precise estimation of some or all of the constants in a mechanistic model. *Can. Journ. Chem. Eng.*, **47**, 76–80.

KARLIN, S. and STUDDEN, W.J. (1966). Optimal experimental designs. *Ann. Math. Statist.*, **37**, 783–815.

KIEFER, J. (1958). On the nonrandomized optimality and randomized non-optimality of symmetrical designs. *Ann. Math. Statist.*, **29**, 675–699.

KIEFER, J. (1959). Optimum experimental designs. *J.R. Statist. Soc. B.*, **21**, 272–319.

KIEFER, J. (1974). General equivalence theory for optimum designs (approximate theory). *Ann. Statist.*, **2**, 849–879.

KIEFER, J. and WOLFOWITZ, J. (1960). The equivalence of two extremum problems. *Canad. J. Math.*, **12**, 363–366.

LINDLEY, D.V. (1956). On a measure of the information provided by an experiment. *Ann. Math. Statist.*, **27**, 968–1005.

PEREIRA, B. de B. (1977). Discriminating among separate models: a bibliography. *Int. Statist. Rev.*, **45**, 163–172.

PUKELSHEIM, F. (1979). On c-optimal design measures. Preprint.

PUKELSHEIM, F. (1980). On ϕ_p-optimal design measures: characterizations based on duality theory. Preprint.

ROCKAFELLER, R.T. (1970). *Convex analysis*. Princeton U.P., Princeton N.J.

SEARLE, S.R. (1971). *Linear Models*. New York: Wiley.

SIBSON, R. (1972). Contribution to discussion of 'Results in the theory and construction of D-optimum experimental designs' by H.P. Wynn. *J.R. Statist. Soc. B*, **34**, 181–183.

SIBSON, R. (1974). D_A-optimality and duality. Progress in Statistics. *Colloq. Math. Soc. Janos Bolyai*, **9**, 677–692.

SILVEY, S.D. (1972). Contribution to discussion of 'Results in the theory and construction of D-optimum experimental designs' by H.P. Wynn. *J.R. Statist. Soc. B*, **34**, 174–175.

SILVEY, S.D. (1978). Optimal design measures with singular information matrices. *Biometrika*, **65**, 553–559.

SILVEY, S.D. and TITTERINGTON, D.M. (1973). A geometric approach to optimal design theory. *Biometrika*, **60**, 21–32.

SILVEY, S.D., TITTERINGTON, D.M. and TORSNEY, B. (1978). An algorithm for optimal designs on a finite design space. *Commun. Statist. A7*, **14**, 1379–1389.

SMITH, K. (1918). On the standard deviations of adjusted and interpolated values of an observed polynomial function and its constants and the guidance they give towards a proper choice of the distribution of observations. *Biometrika*, **12**, 1–85.

TSAY, J.-Y. (1976). On the sequential construction of D-optimal designs. *J. Amer. Statist. Assoc.*, **71**, 671–674.

VUCHKOV, I.N. (1977). A ridge-type procedure for design of experiments. *Biometrika*, **64**, 147–150.

WHITE, L. (1975). Ph.D. Thesis. Imperial College, London.

WHITTLE, P. (1971). *Optimization Under Constraints*. New York: Wiley-Interscience.

WHITTLE, P. (1973). Some general points in the theory of optimal experimental design. *J.R. Statist. Soc. B*, **35**, 123–130.

WU, C.-F. (1978). Some iterative procedures for generating nonsingular optimal designs. *Commun. Statist. A7*, **14**, 1399–1412.

WYNN, H.P. (1970). The sequential generation of D-optimal experimental designs. *Ann. Math. Statist.*, **41**, 1655–1664.

WYNN, H.P. (1972). Results in the theory and construction of D-optimum experimental designs. *J.R. Statist. Soc. B*, **34**, 133–147.

INDEX

Algorithms, 28–39
  convergence of, 35–36
  exchange, 34, 39
  V-, 30
  W-, 30, 32, 34
Approximate theory, 13–52
  for c-optimality, 49
    D-optimality, 40–43
    D_A-optimality, 43–44
    D_s-optimality, 45–48
    linear criteria, 48–49
Ash, A., 2
Atwood, C.L., 1, 11, 34

Bandemer, H., 2
Bayesian inference, 66
Box, G.E.P., 1
Brooks, R.J., 2

Carathéodory's theorem, 16, 72–73
Chernoff, H., 1, 68
Classical design, 5–6
Concave function, 17, 75–77
Concavity results, 69–71
Confidence ellipsoid, 10
Confidence intervals, 66–68
Consistency, 66
Control variables, 2, 4
Convex
  combination, 16
  cone, 72, 76
  hull, 16, 49, 54, 72–73
  set of information matrices, 16
c-optimality, 13, 32, 49
Criterion function, 12, 14, 16
Cylinder. *See* Thinnest s-cylinder

Decision theory, 2
Design
  criteria, 10–13
  measure, definition of, 15
  space, definition of, 6
Differentiability, 18, 74–77
Directional derivatives, 17–19, 74–75
Discrimination among models, 8
D-optimality, 10, 21, 28, 40, 79
D_A-optimality, 10, 43
D_s-optimality, 10, 45
Draper, N.R., 1
Duality, 27, 41, 46, 78, 81

Efficiency, 58, 60, 61
Elfving, G., 1, 13, 49, 51, 58
Entropy, 59
Envelope
  lower, 71
  upper, 70
E-optimality, 12, 25
Equivalence theorem, 22, 23
Estimator
  best linear unbiased, 5
  least-squares, 5
  maximum-likelihood, 4, 65
  unbiased, 3

Fedorov, V.V., 1, 12, 29, 30, 32, 35, 38, 39, 43, 63
Fisher's information matrix, 3, 7, 53, 55, 62, 65
Ford, I., 59, 64, 65
Fréchet derivative, 18, 44, 48, 75

de la Garza, A., 1
Gâteaux derivative, 17, 44, 48, 74
Generalised inverse, 25, 27, 44, 69

INDEX

G-optimality, 12, 23, 41
Gradient, 74
Guest, P.G., 1

Hedayat, A., 2
Henson, T.L., 1
Hill, W.J., 1
Hoel, P.G., 1
Hunter, W.G., 1

Induced design space, 9, 15, 20, 32
Inference. *See* Bayesian; Likelihood; Repeated-sampling; Sequential designs
Information, 4
Information matrix
  definition of, 15
  singular, 25, 51, 56, 58, 64
  See also Fisher's information matrix

Karlin, S., 1, 11, 46
Kiefer, J., 1, 12, 17, 23, 27, 52, 69

Lagrange multipliers, 78
Lagrangian principles, 78
Lagrangian theory, 51, 78
Least-squares, 5, 9
Legendre polynomial, 43
Likelihood inference, 66
Lindley, D.V., 59
Linear criterion function, 12, 48
Linear regression. *See* Regression
Linear theory, definition of, 9
Logistic quantal response model, 59

Malyutov, M.B., 1, 63
Maximin designs, 59, 61
Maximum-likelihood. *See* Estimator
Minimum ellipsoid, 41, 78, 79

N-observation design, 2, 9, 37–39
Non-differentiability, 24, 25, 37
Non-linear design, 6, 7, 53–68

Pereira, B. de B., 9
ϕ-optimality, definition of, 15
ϕ_θ-optimality, definition of, 54

Polynomial regression, 42
Prior distribution, 2, 59
Probit analysis, 7
Pukelsheim, F., 17, 26, 27, 46, 49, 52, 78, 81

Regression
  linear, 4, 15
  polynomial, 42
  quadratic, 6, 20
Repeated-sampling inference, 63, 66–68
Rockafeller, R.T., 18, 74–76, 78

Searle, S.R., 25, 27
Sequential designs, 62–68
  convergence of, 64
  inference from, 66
Sequential experimentation, 8
Sibson, R., 11, 41, 46, 71, 78, 80
Silvey, S.D., 11, 26, 27, 28, 34, 41, 46, 51, 64, 65
Simulation, 65, 67
Singular information matrix, 25, 51, 56, 58, 65
Slack variables, 79
Smith, K., 1
Step-length, 30
Studden, W.J., 1, 11, 46

Thinnest s-cylinder, 46
Titterington, D.M., 11, 28, 34, 46
Torsney, B., 28
Tsay, J.-Y., 34

V-algorithm. *See* Algorithm
Variance matrix, 3, 4, 10
Vuchkov, I.N., 34

W-algorithm. *See* Algorithm
White, L., 54, 63, 64
Whittle, P., 1, 18, 41, 78
Wilson, K.B., 1
Wolfowitz, J., 12, 23
Wu, C.-F., 28, 34
Wynn, H.P., 1, 28, 29, 30, 33, 35

THE UNIVERSITY OF MICHIGAN

DATE DUE

MAR 0 9 1997

APR 2 1 2000

MAY 2 6 2004

NOV 2 2 2005

NOV 1 2 2007

APR 1 7 2009